MA 3 2410 01990 3635

W9-BGQ-523

Weird Life

ALSO BY DAVID TOOMEY

The New Time Travelers

Stormchasers

WEIRD LIFE

THE SEARCH FOR LIFE THAT IS VERY, VERY DIFFERENT FROM OUR OWN

David Toomey

W. W. NORTON & COMPANY
NEW YORK   LONDON

For inform
v

For

Prologue

One of my favorite books as a child (and, truth be told, one of my favorites today) is Dr. Seuss's *If I Ran the Zoo*. The "I" of the title is Gerald McGrew, a boy of perhaps age ten or twelve. The book opens as young McGrew visits his local zoo and finds its animals—a few sleepy-eyed bears and lions— uninspiring. He wishes for more exotic creatures, and so begins an extended daydream. Our youthful protagonist imagines himself as the zoo's manager, releasing the bears and lions, which shamble off, one presumes, to find more stimulating environs. Then he fantasizes himself the zoo's procurer, outfitted with pith helmet and butterfly net, in the heroic mold of Mallory and Burton, scaling mountains and crossing oceans in search of ever more exotic animals. And he finds them. In the "Desert of Zind" he captures a ferocious sort of camel called a "Mulligatawny," and on the "Island of Gwark" he catches a gigantic bird called a "Fizza-ma-Wizza-ma-Dill." And then he assures us that he's just warming up.

Young Gerald McGrew may be a fictional character in a

children's book, but his urges resonate. Humans, it seems, have always been less than satisfied with actual fauna, and so moved to invent alternatives. We all remember a few: the sphinx, the griffin, the basilisk, the phoenix. But ancient cultures created many more. Margaret Robinson's *Fictitious Beasts* (one of the most thorough and authoritative catalogues) lists several hundred, each description replete with details of behavior and in many cases an instructive encounter with a god or heroic mortal.[1] To a biologist, what is striking about this imaginary bestiary, especially in comparison with the bestiary that nature actually has produced, is its paucity. The fact is that no one knows exactly how many species reside on our planet at present, but a conservative guess is 3.6 million, and some estimates are as high as 100 million.[2] To those who prefer the real to the imaginary, it should come as good news that Robinson's work has a real-world cognate. The *Encyclopedia of Life*, an effort to build an online compendium of every extant species, is now at 500,000 web pages and growing.

Peruse Robinson's work at a rate of a second per page, and you'll close the back cover after about four minutes. To get through the *Encyclopedia of Life* at the same rate you'll need six weeks. Even then, you'll have only a hint of the astonishing diversity of life over time. There have been at least *30 billion* distinct species in the history of Earth.[3] Suppose there were a book devoting a page to each species. To read it at a rate of a second per page, you'd need nearly ten centuries.

It's not just the numbers. By comparison with reality, the creatures of myth suffer in another way. Most are little more than portmanteaus—with, for instance, the head of one animal sewn onto the body of another. Any imaginatively sadistic school-child equipped with scissors and glue, you might think, could do as well. Even the parts list is severely scaled back, derived, as it is, almost entirely from the single branch of the tree of life that bears mammals, lizards, and birds. There are exceptions like "Grandmother Spider," who figures in many Native American creation myths; and the kraken, the giant squid that inhabits the imaginations of coastal dwellers on several continents. And there are some rather wondrous hybrids. The "vegetable lamb of Tartary" (*Agnus scythicus* or *Planta Tartarica Barometz*), for instance, was a legendary plant of central Asia believed to grow sheep as its fruit. The sheep were connected to the plant by an umbilical cord, and when they had eaten all they could reach, the whole plant withered and died.[4] But that is about as bizarre as the mythical beasts get. The overwhelming majority, if we were to classify them within standard *taxonomies*, would be *vertebrates* in the phylum Chordata—essentially, animals with backbones.

For most of human history, anyone seeking a little strangeness in the proportion had to be satisfied with these. Then, in the mid-seventeenth century, natural philosophers discovered another bestiary, one that had two advantages over the imagined one. First, its animals were real. Second, they did not inhabit an exotic and distant country. In fact, they lived among us. And on us. A great many of them lived *inside* us.

By the late 1620s, the first of the type of instrument we call the microscope had been crafted and named. Half a century later, a twenty-eight-year-old British naturalist named Robert Hooke began to make his own, to use them to examine a great many things, and to sketch what he saw. Hooke's work was not always easy. He had two sorts of microscopes. One, rather like the instrument we know, was a set of lenses fixed and aligned inside a small tube; the other, far more difficult to use, was a glass bead the size of a pinhead, held in a brass mounting. Of course, the subjects themselves could be uncooperative. The only way Hooke could immobilize ants without half crushing them was to get them drunk on brandy.

Despite such challenges, by 1664 Hooke had produced a work called *Micrographia*. Even now, in an age of high-definition television and IMAX 3D, it can be a startling experience to open the book to a folded page and gently pull its leaves apart to reveal, for instance, a copperplate engraving of a flea measuring a monstrous half meter across. One of Hooke's contemporaries, Samuel Pepys, called the book the "most ingenious" he had ever read. *Micrographia* became a best seller, and many of its readers would have agreed with Hooke's observation of the flea: "the strength and beauty of this small creature, had it no other relation at all to man, would deserve a description."[5] Still, a few criticized Hooke's pursuit of knowledge with no obvious practical application, and many more were simply uninterested. The reason may have been that the animals discovered by Hooke and his successors were utterly unlike anything known at the time; they were perhaps *too* strange. Of course, there was another reason for the disinterest: Hooke's creatures were very, very small. Then, as now, humans equate small with unimportant—a prejudice that, as we shall see, is as misguided as it is dangerous.

Some two centuries after *Micrographia*, natural philosophers discovered another bestiary—this one populated by animals that were as large by comparison with us as we are compared to house cats. They are long vanished from the Earth, but their appeal, perhaps especially to a certain demographic of seven-year-olds, has proved enduring. Some of the reasons are obvious. Dinosaurs are strange enough to inspire wonder, but not so strange as to be wholly unfamiliar. *Tyrannosaurus rex* (a favorite of the aforementioned demographic) saw through eyes not unlike our own, breathed through nostrils, and walked over ground.*

As the flea and *Tyrannosaurus rex* demonstrate, nature itself will outperform the uninformed imagination every time. If these creatures showed the limits of human imagination, they also enlarged it. With the discovery of microscopic and sub-microscopic life came questions about nourishment, reproduction, and mobility. Was there cooperation? How small was it possible for a living thing to be? The dinosaurs, likewise, produced new questions. How could such enormous masses be supported? How large was it possible for a living thing to be? And of course, why did they vanish?

Even as naturalists pondered these questions, there came a new understanding of life at fundamental levels. In 1859, Charles Darwin, prodded by the independent discoveries of British naturalist Alfred Russel Wallace, published *On the Origin of Species by Means of Natural Selection*. Despite attacks by religious conservatives, the book was widely read (it would see six editions in Darwin's lifetime), and within a decade several works were published in its support. In the early years of the twentieth century, biologists rediscovered Gregor Mendel's laws of heredity, and American geneticist Walter Sutton found evidence that *chromosomes* carry units of inheritance. In the 1940s, Julian Huxley and George Gaylord Simpson consolidated natural selection and genetics, and *DNA* was found responsible for hereditary changes

---

* In the popular imagination, dinosaurs were real enough that Dickens, in *Bleak House*, could use a resident of the Middle Jurassic to add a touch of atmosphere to a rainy day on the outskirts of London. "As much mud in the streets, as if the waters had but newly retired from the face of the earth," he wrote, "and it would not be wonderful to meet a Megalosaurus, forty feet long or so, waddling like an elephantine lizard up Holborn Hill." (p. 1)

in bacteria. In 1953, James Watson and Francis Crick published the structure of DNA in the journal *Nature*.

Meanwhile, as to the variety of forms that might be made with that DNA, there were—to the delight of most and the consternation of some—more surprises. Again and again, scientists discovered animals and plants that broke all the rules, surpassing what many had assumed to be limits in size, shape, and behavior. Nonetheless, by the mid-twentieth century most biologists had reason to believe that life would survive within only a narrow range of pressures and temperatures. There seemed to be some limits that could not be surpassed.

By this time there were at least nine specialties in *biology*, the study of living organisms and life processes. Probably it shouldn't be surprising that practitioners in each specialty tended to define life in the terms of that specialty, and that they had no shared definition of the core subject at all. But what *is* surprising is that no one thought this lack of consensus much of a problem. Taxonomists, molecular biologists, and embryologists went about their business identifying species, studying chemical reactions that maintained a *metabolism*, and culturing *microbes*. If asked to define life by, say, an upstart philosophy major at an interdepartmental faculty reception, they would say they knew it when they saw it and that, thank you, was quite enough.

In the early 1970s, however, it became obvious that the biologists' confidence in their powers of recognition, whether justified or unjustified, was not quite enough. It was about that time that NASA asked scientists to submit designs for life-detecting experiments to be carried to the planet Mars aboard the two unmanned *Viking* spacecraft. These would be the first *in situ* attempts to discover life on another world. To detect something with a miniature laboratory that would be operated remotely by a radio signal sent from a transmitter more than a million miles away was no small challenge. It seemed reasonable that the task would be made easier if the "something" to be detected was properly defined first.

The three experiments chosen by NASA were ingenious but, at least in the view of some, lacking in imagination. Two were designed with the assumption that Martian life would need water, and all three were designed under the assumption that life would survive in only a narrow (in fact, a rather Earthlike) range of temperatures. Depending on whom you ask, the results of that reconnaissance meant that either there was no life in the spacecrafts' vicinity or (this from one experiment) there might be some very unusual life indeed. In any case, the results did little to change larger ideas of life's boundaries.

Then, in a series of discoveries in the 1980s and 1990s back on Earth, scientists found that they had underestimated nature's ingenuity; the realm of life was (again) greater than they had dared imagine. In places where no one thought life possible, organisms were not merely surviving; they were thriving. Once biologists began to look, they found them everywhere. And there were enough to satisfy an army of Gerald McGrews.

No one expected life in water much above its boiling point. But scientists found bacteria living in volcanic hydrothermal vents on the ocean floor, one species merrily reproducing at a scalding 235°F. No one thought life could survive in water at temperatures much below its freezing point. But in Antarctic ice floes, scientists found channels of slushy brine in which single-celled algae were harvesting energy from the sunlight filtered through ice and assimilating nutrients from the water below.

Biologists had assumed other limits as well. They had thought that organisms would tolerate only a narrow range of pH levels.

Then they discovered life flourishing in hot sulfur springs and growing vigorously in soda lakes. They had assumed that aquatic life would tolerate only so much salt. But they found bacteria that had adapted perfectly to saturated salt lakes. They had believed that high levels of radiation would kill any organism. But they discovered a bacterium that, by efficiently repairing broken DNA strands, could withstand radiation energies at a thousand times the level that would kill a human. They had assumed that life required a "substrate"—a surface on which its molecules could interact easily and often. But they found microbes that may be living through their entire life cycles—growing, metabolizing, reproducing—in clouds.[6]

Biologists knew that many creatures living in the dark on the seafloor ate organic material that fell slowly from the surface, and that some survived by drawing energy from chemical reactions. Still, they assumed that all life depended—perhaps indirectly but nonetheless ultimately—on the Sun. But in 1996, a group of scientists reported the discovery, more than a mile beneath Earth's surface, of assemblages of bacteria and fungi that gained all their energy from inorganic chemicals in the rock around them.

All these organisms—the bacteria in the hydrothermal vent, the algae in the Antarctic brine, the rock-eating fungi, and the rest—came to be known collectively as *extremophiles*—lovers of extremes.

The physical boundaries within which life is possible are unknown and undefined, but most biologists believe that they must exist, for the simple reason that there are temperatures and pressures under which the structures of organisms—cells, DNA, and proteins—will break down, no matter how well protected. In short, life must have *ultimate* limits. If life exists outside them, it must be something other than what has been called, in that venerable phrase that can hardly be improved upon, "life as we know it." It must be fundamentally different. It must be, in a word, *weird*.

Exactly what can we say about such life? At the very least, we can say what it is not. All life we know has DNA, the same twenty or so amino acids and proteins, and a biochemistry that employs the same thousands of chemical pathways (the complex chemical reactions by which a metabolism is maintained) and that uses liquid water as a solvent. It is for these reasons that biologists believe that all life we know—you, me, the flea, the megalosaurus, Charles Darwin, your neighborhood creationist, and all the extremophiles—is descended from a single common ancestor. Weird life, if it exists, would *not* be descended from that ancestor, and it could be weird in any number of ways. It might have as its basis a molecule other than DNA, it might use other amino acids, or it might use a solvent like ammonia or liquid methane.

Whether such substitutions are possible, or whether the fundamental features of life we know are necessary to all life, is far from clear. But it may be that some, or many, or most of those features are the products of happenstance, and that life on Earth might as easily have taken a different path, the result being organisms whose fundamental features would be utterly different from those we know. What *is* clear is that the discovery of even one example of such life would profoundly change our understanding of biology. The familiar illustration of all life we know is a great tree, its trunk splitting and splitting again into branches representing phylogenetic categories, each less fundamental and more populous than that from which it sprouted, finally ending in millions of twigs representing individual species. There have been many discoveries of new species, and recently, even a new phylum.[7] But an example of another sort of life would mean that the tree *itself* is not unique, and that it may be only one of many, perhaps one in an entire forest. Such a discovery would be cause for humility on our part—another demonstration that we occupy a smaller part of the universe than we now believe. It would also be cause for renewed wonder at a cosmos that is stranger and vastly richer than we now imagine.

Papers on what we now call weird life appeared sixty years ago, but they were few and scattered across disciplines. There was no overview and nothing like an ongoing discussion. The first gesture in those directions came in 2002. NASA and the European Space Agency (ESA) were thinking about (if not exactly planning) unmanned missions to the outer Solar System—to Saturn's moon Titan, to Neptune's moon Triton, to the comets. If life existed in these places, it would be radically different from anything we knew, and radically different from anything that *Viking* looked for on Mars.

The idea is not new, and we've seen it again and again in science fiction: An astronaut on a desolate but otherwise unremarkable planet sees what he assumes is an odd rock formation. He looks away, and it moves. He looks back and realizes his mistake, and (as they say) dramatic complications ensue. But this was not science fiction. Now NASA and ESA were taking the possibility of weird life quite seriously, and making it clear that much was at stake. The discovery of extraterrestrial life—whether it developed independently or had migrated from planet to planet via meteor, solar wind, or some other means—would have profound and lasting consequences not merely for the life sciences or even for science in general, but for our understanding of our very place in the universe. But would we recognize life if we saw it? And if we did *not* recognize it, might we, by inattention or carelessness, destroy it?

In 2002 the National Research Council (NRC) assembled a group of twenty-five scientists from research laboratories and institutes across the United States. The group called itself the Committee on the Limits of Organic Life in Planetary Systems, and its task was nothing if not ambitious. Its members were to define life, to identify the traits necessary to life as we know it, and to determine the outer limits of living systems. As if this were not enough, they were handed a second, more provocative challenge: to imagine possibilities for weird life in some detail. For five years they read and discussed papers, collected data, and talked. By summer 2007 they had published a report summarizing their work. It was titled *The Limits of Organic Life in Planetary Systems.*[8] Publication of the NRC report represented a watershed moment in the history of thinking about the boundaries of life as we know it, and what sort of life might lie beyond those boundaries. It also provides much of the foundation for this book.

A word on nomenclature. New fields of study try on many names, and this one is no exception. The subject of this book might, with varying degrees of justification, be called beta life, hypothetical life, nonstandard life, nonterran life, unfamiliar life, life as we do *not* know it, alternative biology, and (you knew this was coming) Life 2.0. I've settled on *weird life* because at the time this book goes to press it seems to enjoy the most widespread usage, and because it conveys with great economy the sense of strangeness the subject deserves.

As we'll see, in the violent early history of the Solar System, Earth and Mars traded material, some of which may have been biological. A few scientists argue that life we know may have emerged from spores of extraterrestrial origin. For the time being I will sidestep the question of place of origin, and unless otherwise noted, I'll use the term "familiar life" to mean all life—

terrestrial and otherwise—that is descended from the single ancestor of all life we know, allowing for the possibility that that ancestor may have been extraterrestrial. Alternately, I will use "weird life" to mean any organism or organisms—again terrestrial and otherwise—that are *not* descended from that ancestor.

Weird Life

CHAPTER ONE

Extremophiles

Julie Huber is a marine oceanographer working at the Marine Biological Laboratory in Woods Hole, Massachusetts. She is thirty-four but has an easy laugh that makes her seem even younger, and if you saw her jogging on a beach you might take her for a professional volleyball player. Her accomplishments in the field of oceanography and marine biology are many; she has logged nearly a year on oceangoing research vessels, and made several descents in the most famous of all research submarines, *Alvin*. Her main research focuses on the microbes that live in and beneath the crust of the seafloor. They are organisms that sequester carbon, cycle chemicals, and affect the circulation of ocean water—all of which are activities crucial to the oceans' overall health. Yet these same organisms happen to be, in the language favored by field biologists, "vastly undersampled." Consequently, they have been little studied and so are little understood. They are also in places so difficult to reach that if Huber hopes to study them she must use techniques from various disciplines, among them geology, genetics, and molecular chemistry.

Dr. Huber is possessed of a passionate intellect that can catch you by surprise. She will look you straight in the eye and affirm that a certain development in microbiology "really excites the marine sediment community."[1] While you register the fact that there *is* a marine sediment community and consider that its members could perhaps benefit from some excitement, she is already explaining, in language that is an unself-conscious mash-up of technical and colloquial, the particular challenges of trying to detect organisms by measuring the chemical components of seawater, noting by the way that certain recent work on methanogen diversity is very cool.[2]

Lately Huber has been, in her words, "chasing seafloor eruptions." She is particularly interested in organisms that live in the Pacific, near three "seamounts" (oceanographers' term for submarine mountains). All three seamounts are geologically active, and Huber makes periodic visits to each—or, more often, to the place on the ocean's surface several kilometers above them. The visits usually hold surprises. In May 2009 she was on a research vessel in the western Pacific, only twelve hours out of port in Samoa, when the remote operating vehicle *Jason*, from 2 kilometers beneath them, began transmitting live, full-color video images of the West Mata volcano erupting. It was 3:00 a.m. on the ship, but everyone—scientists and off-duty crew alike—was crowded into one room watching the televised images of lava flowing, sometimes explosively, from the deepest erupting volcano anyone had ever seen. It was, Huber said, "definitely the coolest thing I've seen on the seafloor," adding wryly, "and I've seen a lot of seafloor."[3]

The water in the samples that *Jason* retrieved from the vicinity of the volcano was as acidic as battery acid, yet it contained living bacteria. There were fewer kinds than at similar sites, and so there was a less diverse *microbial community*—this the phrase used to describe many populations of microbes living together, sharing resources, and in various ways making life better for one another. Whether the relative paucity of kinds means the environment is too harsh for certain organisms, or whether West Mata's young age means that things are just getting started there, is an interesting and open question—one that Huber hopes to answer in the months ahead.

At the moment, Huber is managing several projects simultaneously. Late on a Friday morning in March 2010, she has just received an e-mail from the postdoctoral student studying under her and doing fieldwork on a research vessel near Guam. It seems that they had set markers and moorings on the seafloor and now couldn't find them. Huber gives a "this sort of thing happens all the time" shrug and suspects the culprit is what geologists, rather unromantically, term a "slumping event." In all likelihood a part of the seafloor slid sideways, taking the markers and moorings with it. Just another reminder, not that Huber needed reminding, that the seafloor is not the grave-quiet place that only half a century ago, many scientists believed it to be.

Huber's research may trace its origins, quite directly, to the discovery in 1977 of the so-called hydrothermal vent communities, and less directly to questions that arose in the first decades of the twentieth century—as to why the continents are shaped as they are.

A SCIENTIFIC MYSTERY

Anyone who sees a world map centered on the Atlantic Ocean cannot help but notice that the east coast of South America seems made to fit, jigsaw puzzle–like, into the inward-bending coast of Africa. In 1922 a German geologist and meteorologist named Alfred Wegner went several steps further. Assembling the evidence of fossils, mineral deposits, and scars left by glaciers, he proposed that the comparison was apt. The continents *were* pieces of a puzzle, pieces that happened to be slowly drifting apart. In the decades that followed, others developed Wegner's hypothesis into a theory of plate tectonics, which proposed that the Earth's crust is composed of plates—perhaps ten "major" ones and as many as thirty "minor" ones. It was thought that their upper parts were brittle, their lower parts warmer and more malleable, and that some might be as much as 80 kilometers thick. Geologists found evidence that molten rock was pushing into the seams where the plates had pulled apart.

If this were the case, it might help to answer a question that was surprisingly long-standing and surprisingly straightforward: Why is the chemistry of seawater what it is? Lakes like the Dead Sea—lakes with no outlet other than evaporation—are called "closed basins." They are alkaline in the extreme, and they grow more alkaline over time. Logically, since the world's oceans have no outlet, like very large closed-basin lakes, they should be very, very alkaline. Yet their pH, the measure of the acidity or alkalinity of a solution, was between 7.5 and 8—very near the middle of the scale—and this was the case for foaming breakers in the Florida Keys; for dark, dense water in the Mariana Trench; and for the frigid water lapping Antarctic icebergs. It seemed clear that some process was at work, filtering the water and maintaining the pH, and doing it everywhere. A few scientists began to suspect undersea hot springs, and they had ideas as to their whereabouts. Hot springs on the Earth's surface were heated by the molten rock in nearby volcanoes; it seemed reasonable to expect that hot springs on the seafloor would also be near molten rock. And many thought there was molten rock in the seams between tectonic plates. Find the seams, many suspected, and you'd find the hot springs.

But no one knew for sure. Even by the early 1970s, textbooks in oceanography introduced their subject with the startling fact that we knew less about the ocean floor than we knew about the near side of the Moon. If anything, this appraisal was generous. Sonar for mapping the floor was crude, and equipment used to measure temperatures and pressures was towed on cables behind ships. Sooner or later a cable would snag on an undersea rise, and the ship would idle its engines while a dispirited crew pulled the equipment aboard. It usually came back wrecked—unless the cable just broke, in which case it didn't come back at all.

The United States Navy, however, had developed sophisticated techniques for mapping the ocean, and by the mid-1970s the navy had begun to share them with researchers. Using these techniques, scientists at Woods Hole Oceanographic Institution (WHOI) implemented a three-stage method to explore a swath of seafloor. First, the research ship *Knorr* would drop transponders. Because the seafloor is uneven, they would settle at different depths. Then their positions were measured with great precision by sonar, allowing researchers to derive a low-resolution map of the terrain. Finally, a camera vehicle—a 1.5-ton "gorilla cage" mounted with cameras, strobe lights, and power supplies—would be towed over the terrain at a cautious 4 kilometers an hour, 20 meters above the seafloor. Every few hours the crew would haul the vehicle aboard, pull the film, and develop it.

In the spring of 1977, Woods Hole researchers on the *Knorr* were mapping the seafloor in the eastern Pacific about 280 kilometers northeast of the Galápagos Islands. After one run over terrain 2 kilometers deep, the film showed white clams. It was obvious they were alive.

At such moments the research submarine *Alvin* (owned by the United States Navy and operated by WHOI) would be called into play. By the 1970s she was already something of a workhorse. In 1968 she had been lost once and recovered. Two years earlier, when an air force B-52 bomber collided with a tanker over the Mediterranean Sea and (accidentally) dropped an undetonated hydrogen bomb, *Alvin* was given its moment in Cold War history and summoned to search the ocean floor off Spain. On March 17, 1966, *Alvin*'s pilots found the bomb resting on the seafloor nearly 910 meters deep. It was raised intact. By 1977 *Alvin* had had several upgrades, but its fastest speed was a modest 4 kilometers an hour, and its lights penetrated only 15 meters. Not that any of this mattered. What *Alvin* was especially good for, and what she is still good for, is close observation. And so, when the *Knorr* found live clams 2 kilometers beneath the surface, *Alvin* was towed to the site.

Alvin's crew compartment, a hollow titanium sphere 3 meters across, holds three—a pilot and two researchers. The researchers for this particular investigation were geologists—John Corliss of Oregon State University and John Edmond of MIT. They spent most of the 2,000-meter descent peering out the Plexiglas portholes. There wasn't much to see, and even when they were a few meters above a sloping seafloor, *Alvin*'s lights illuminated nothing but the hardened molten rock that geologists call pillow basalt. It covered the sloping floor in all directions, making for a scene that, even to a geologist, was not particularly remarkable. Then they noticed something about the water itself. It was shimmering like the air over a hot grill. Hurriedly, Corliss and Edmond took measurements and found that the water was warmer than water at this depth should be, by about 4 degrees.

The pilot took *Alvin* up the slope, and when they neared the crest of a ridge they were astonished to see, lit by the searchlight and through the shimmering water, reefs of mussels, giant clams, crabs, anemones, and fish. It was a fantastic undersea garden, an oasis vibrant with life. Corliss and Edmond did not know how the riotous island *ecosystem* around them was possible. They did know, however, that *Alvin* had only five hours of power remaining, and they spent that time, Edmond would later write, "in something close to a frenzy," measuring water temperature, conductivity, pH, and oxygen content, and taking samples of everything that *Alvin*'s mechanical arm could grab.[4] That evening, back aboard the *Knorr*, there was a small celebration. Someone had a camera and snapped photos of Corliss and Edmond—young men, bleary-eyed and smiling.

In 1830, British naturalist Edward Forbes claimed that because sunlight could not penetrate deeper than 600 meters, phytoplankton could not survive below that depth. Without phytoplankton, there was no base for a food chain. It followed, reasonably enough, that the deep ocean must be sterile.[5] By the mid-twentieth century, the processes by which oceanic life sustains itself were well understood. Sunlight supplies the energy. Nutrients in the form of nitrogen and phosphorus are brought in by rivers and streams and stirred up from the seafloor by upwelling currents. The floating single-celled plants called phytoplankton use the sunlight, nutrients, and carbon dioxide dissolved in the water. They are eaten by the tiny invertebrates called zooplankton that float freely throughout the seas and other bodies of water, and the zooplankton are eaten by shrimp and other crustaceans, all the way up the food chain to the braised tuna with lemon on your dinner plate. Obviously, such processes could operate only near the ocean surface.

In the decades that followed, however, scientists came to realize that there *was* life at great depths. Fish, crabs and other organisms lived in total darkness at enormous pressures, and survived by feeding on dead and decaying matter that sank slowly from the waters above. By the mid-twentieth century, advances in nautical engineering allowed biologists to see this life firsthand. In the mid-1940s the Swiss scientist Auguste Piccard designed a vessel he called a "bathyscaphe." Unlike its predecessor the bathysphere, a simple spherical pressure chamber lowered and raised by a cable, Piccard's new design featured a float chamber for buoyancy and a separate pressure sphere for the crew. The third bathyscaphe Piccard built was called the *Trieste*. It was sold to the United States Navy in 1957, and three years later it took Jacques Piccard (Auguste's son) and navy Lieutenant Don Walsh to the bottom of an undersea canyon called the Mariana Trench. It was there that they noticed, more than 11 kilometers beneath the surface, the deepest place in any ocean on Earth, a flatfish.[6]

Still, even in 1977, most marine biologists expected such organisms would be few and solitary. And since recycling of decaying matter in the ocean's upper levels is fairly efficient and allows very little to sink much lower before it is consumed, they expected those organisms to be quite hungry. So in 1977, when the Woods Hole expedition's chief scientist called a marine biologist named Holger Jannasch to give him the news of a thriving community of life 2 kilometers deep, Jannasch simply didn't believe him. "He was," Jannasch explained, "a geologist, after all."[7]

The expedition would conduct fourteen more descents to the site. It became apparent that Corliss and Edmond had happened upon a hot-spring field. Warm water was flowing up through every crack and fissure in a roughly circular patch of seafloor about 100 meters across. While *Alvin* investigated the newfound life below, scientists aboard the *Knorr* studied the water samples already returned, and found that all had a high concentration of hydrogen sulfide. That turned out to be a thread that wove together an entire *ecology*.

On land, some bacteria were known to derive energy from hydrogen sulfide through a process called chemosynthesis. They were rare, and most organisms took their energy, directly or indirectly, through photosynthesis. But in the dark 2 kilometers deep, chemosynthesis might be the only synthesis possible. Soon, researchers at Woods Hole developed a model to describe the process. It was this: Deep within the Earth, naturally radioactive materials produce heat that melts rock into the substance called magma. Magma is pushed up through the seams between the midocean ridges, where it cools and spreads outward to become new oceanic crust. Meanwhile, seawater continually percolates down through the crust, where the sulfate it carries combines with iron in the crust to produce hydrogen sulfide and iron oxides. When the same seawater, now heated, is pushed back up through cracks and fissures in the crust and returned to the deep ocean, it carries hydrogen sulfide that certain bacteria find quite tasty. The same bacteria absorb oxygen dissolved in the water, and some of that oxygen combines with sulfite to become sulfate.[8]

We would seem to have come full circle, returning to the

chemistry we began with. But the story is not quite over. As you may recall from chemistry class, some reactions absorb energy, while others release it. The chemical reaction that yields sulfate releases energy—which the bacteria, in lieu of sunlight and in a model of efficiency—use to drive their metabolism. From here on up, the food chain of what would come to be called hydrothermal vent communities was thought to be, roughly speaking, like that in the sunlit waters 2 kilometers above.

Corliss and Edmond understood that the water issuing from the vents was probably much diluted as it rose tens of meters through the crust, and that the real action, geochemically speaking, must be in the crust, a kilometer or two deeper down. But they would never see or study that chemistry as it was happening. Or so they thought.

Two years later, researchers who were using *Alvin* to investigate warm upwelling on the Pacific Ocean floor near the Gulf of California happened upon its source: natural chimneys of sulfide minerals, 2–3 meters high, furiously pumping water black with iron sulfide and very, very hot. Soon Corliss and Edmond arrived on-site, took their turn in *Alvin*, and measured the temperature of the water released by the chimneys. It was a nearly incredible 300°C. Under an atmospheric pressure at sea level, if you try to heat water gradually, it will boil away long before it reaches that temperature; and if you heat it rapidly to that temperature, it will boil explosively like (in fact, exactly like) a geyser. It is the pressure of 2 kilometers of water that keeps the chimney's water well behaved.

For Woods Hole scientists, the heat presented some challenges. They had to design and build water samplers that would work at high temperatures, and they had to be careful to keep *Alvin* a safe distance from the chimneys, as the heat might soften its Plexiglas portholes enough to implode them. But the work was exciting and welcome. In the months and years that followed, scientists from many different institutes and universities found more vents and more chimneys (they would come to be called "black smokers") along other midocean ridges, and near all of them, a great many living organisms.[*]

The theory of plate tectonics had predicted hot springs in the seams between tectonic plates. In the most dramatic fashion, Corliss and Edmond's discovery of the hydrothermal vents went a long way to support that theory, and thus closed a chapter in geology. At the same time, their discovery of life that was fed and energized by hydrothermal vents opened a new chapter in biology. Like all good chapters, it provoked questions. Exactly what sorts of organisms live in these places, and in what numbers? How did they adapt to the pressure, the dark, the heat? And how exactly did they get there to begin with?

THERMOPHILES AND HYPERTHERMOPHILES

Some of these questions had been answered several years earlier by a microbiologist named Thomas Dale Brock. Brock was an assistant professor at Indiana University, developing an inter-

---

[*] A few years after Corliss and Edmond discovered the community near the Galápagos, scientists returning to the site and conducting a more thorough reconnaissance discovered rows of slender white tubes with red filaments emerging from their tips—organisms now called tube worms, or *Riftia pachyptila*. In 2002 another expedition visited the site—informally termed the "Rose Garden"—and found that it had been covered with hardened lava. The midocean ridge giveth, and the midocean ridge taketh away. And, it seems, giveth again. The same expedition found tiny tube worms and mussels the size of a fingernail—in a place they called, naturally enough, "Rosebud." (Nevalla, "On the Seafloor")

est in microbial ecology—the study of the relationship of micro-organisms with one another and with their environment. In the summer of 1964, on a brief sabbatical, he was among the thousands of tourists visiting Yellowstone National Park. Brock was captivated not by the bison and grizzly bears so much as by somewhat smaller organisms. He noticed distinct colors in the outflow channels of the hot springs, and when he took a closer look he was astonished to see what he later described as "pink gelatinous masses of material, obviously biological."[9]

The water was decidedly hot. In, fact it was nearly boiling. People had seen the pink stuff before, of course, but they did not know what Brock knew: that no microbiologist expected any microbe could live in water this hot. Microbes that live in water at temperatures between 60°C and 80°C are called "thermophiles," and microbes that live in water with a temperature of 80°C or higher are called "hyperthermophiles." But that is now. In 1964, few would have believed hyperthermophiles possible, and a standard textbook recommended that researchers incubate thermophilic bacteria at a temperature of 55°C or 60°C.[10] The temperature Brock measured in the outflow channels was 90°C. For years, Brock had suspected that research limited to lab-grown bacteria would lead to a blinkered view, and here was his vindication. Because no one had thought bacteria could survive at temperatures much higher than 60°C, no one had bothered to look for them.

Exactly how something in plain sight might pass unnoticed by researchers was a very good question. One answer, offered by historian of science Thomas Kuhn, is that people (scientists included) see what they expect to see, and may not see what they don't expect to see. By way of example, Kuhn described an experiment in which subjects were asked to identify the color and suit of playing cards presented to them quickly and in sequence. The experiment used a trick deck. Most of its cards would be found in any deck, but a few were special, with combinations of color and suit, like a red six of spades, that do not appear in a normal deck. The test subjects, shown the cards quickly and in sequence, did not register the special cards as special, and mistakenly assigned them normal combinations of suit and color. When shown the red six of spades, for instance, many saw a red six of hearts. On the second or third run-through, some subjects began to hesitate before answering. On still more run-throughs, several became hopelessly confused, with one nearly unraveling altogether. "It didn't even look like a card that time," he said. "I don't know what color it is or whether it's a spade or a heart. I'm not even sure now what a spade looks like."[11] Only a few recognized the red six of spades *as* a red six of spades. But as soon as they did, they began to look for other special cards. Kuhn made the point that something similar happens in science. When a scientist recognizes something no one else has recognized, Kuhn wrote, there follows "a period in which conceptual categories are adjusted until the initially anomalous becomes the anticipated."[12]

In the fall of 1964, with his own conceptual categories properly adjusted, Brock anticipated the anomalous—and sought it out. He set up a laboratory in West Yellowstone and began to spend his summers exploring the boiling and superheated pools, doing a sort of microbial fishing. At particularly interesting places he would attach one end of a long string to a tree branch and the other end to a glass microscope slide, and drop the slide into the water. A few days later he would retrieve the slide and examine it. On almost every slide he found heavy bacterial growth.

Initially, other microbiologists thought Brock's discoveries too specialized and esoteric to be of wider interest. There were, after all, only so many hot springs in the world, and so there could be only so many species of thermophilic (or hyperthermophilic) bacteria living in them. Then Corliss and Edmond and their many successors found ecosystems whose basis was organisms that thrived in hot water. And the environments that they preferred, while difficult for species like *Homo sapiens* to reach, were not rare. Far from it. Midocean ridges snake along the oceanic crust for tens of thousands of kilometers. By the late 1970s, microbiologists were poring over Brock's published works for ideas on how thermophilic bacteria adapted and how thermophilic ecosystems might work. And since it was easier and a lot cheaper than mounting an expedition to a midocean ridge, quite a few began to visit his old haunts at Yellowstone.

Meanwhile, word of life on midocean ridges was reaching all corners of academia. The life sciences department at most universities and colleges has a large bulletin board mounted outside the department's main office. On that bulletin board one is likely to find a notice of a forthcoming department meeting, announcements of conferences and calls for papers, as well as more personal ephemera, like a scribbled note about lost car keys. Occasionally there is a page torn from a journal on a subject that someone thought might be of general interest. Such was the case in the fall of 1979, when it seemed that every bulletin board outside every main office of every life sciences department at every university and college—and high school too—had an article about the deep-water sulfide chimneys and the life around them.

Soon enough, biologists began to wonder whether there were other "special" cards in the deck, and many began looking. Through the 1980s and 1990s, to anyone reading the science section of a newspaper, it seemed that every other week someone had found life where (one would have thought) it had no reason to be. There were heat lovers, cold lovers, pressure lovers, acid lovers, alkaline lovers, salt lovers, and even radiation lovers.* As a group they became known as *extremophiles*, a term that had been coined by R. D. MacElroy in 1974.[13]

Through much of the twentieth century, biologists classified organisms within a taxonomic system whose largest and most fundamental categories were Animalia and Plantae. Single-celled organisms like bacteria were included among the Plantae, it seemed to some, as an afterthought. In the 1960s, biologists began to regard the system as inadequate, especially with respect to microorganisms, and they developed a new taxonomy in which the most fundamental divisions were five "kingdoms": Animalia, Plantae, Fungi, Bacteria, and Protista. The kingdom of Protista, its boundaries particularly ill defined, included many organisms simply because they fit nowhere else. Certain microbiologists (among them evolutionary biologist Ernst Mayr) proposed a more fundamental division into two "empires." Bacteria, whose cells were relatively small and lacked a nucleus, were classified as *prokaryotes* (*pro* meaning "before" and *karyote* meaning "kernel" or "nucleus"); and the other four kingdoms, whose organisms were composed of larger, nucleated cells, were classified as *eukaryotes* (*eu* meaning "true").

In the 1960s, microbiologist Carl Woese and his colleagues began to sequence ribosomal RNA and realized that many microorganisms that had been classified as bacteria (under a light microscope they *looked like* bacteria) were in fact fundamentally different. The categories were redrawn yet again, this time as three "domains." The eukaryotes were called *Eukarya*, and the

---

* Or, thermophiles and hyperthermophiles, psychrophiles or cryophiles, barophiles or piezophiles, acidophiles, alkaliphiles, halophiles, and radiophiles.

prokaryotes were split into the domain *Bacteria* and the newly discovered domain *Archaea*. Woese's taxonomy is especially pertinent to our interests here. While extremophiles include members of all three domains, most are archaea.

Of course, since they are as unlike each other as they are unlike other life, extremophiles are a group only in the sense that "all composers not Beethoven" or "all painters not Monet" are a group. Any given extremophile can be represented by a different outlying point on a bell curve, and there are bell curves for temperature, pressure, and pH. Many, like a certain species of *Acidianus* that thrives at high temperatures and low pH levels, can be represented by outlying points on *two* bell curves.[14] What counts as extreme, of course, depends on who is ringing the bell. R. D. MacElroy, presumably, had a body temperature of 98.6°F and most probably a distaste for strong acids. If the *Acidianus* species were to categorize him, it would call him a "psychrophile" and an "alkaliphile"—a cold lover and an alkaline lover.[15]

In any case, by the 1990s the search for extremophiles had accelerated. NASA, interested in learning how organisms might adapt to harsh environments like the subsurface of Mars, funded numerous research programs—some independently, some with the National Science Foundation. In 1996 a group of biologists convened the first International Conference on Extremophiles. Within a few years, researchers in the new field had established a journal and a professional society, and had published thousands of papers.

One point of agreement in all this research was that if there is a limit, an outer boundary beyond which the most extreme of extremophiles cannot pass, it was probably set by the swish, gurgle, and drip of liquid water. It so happens that every place scientists have found life, they have also found liquid water or evidence of its presence. And almost every place they have found liquid water, they have found life.[16]

WATER

In a list of chemicals arranged by their molecular weight, you would expect water, with a lower weight than oxygen or carbon dioxide, to be a gas at room temperature. In fact, the only reason water is a liquid hereabouts is that its molecule is polarized—the two hydrogen atoms on one side of the oxygen atom holding a slight positive charge, the oxygen atom itself holding a modest negative charge. It is an arrangement that allows water molecules to form bonds that are gentle, yet strong enough to make water bead on glass, to let it be pulled upward through a plant stem, and to endow it with surface tension—that intriguing quality by which molecules on the liquid's surface are attracted to each other more powerfully than they are to the air molecules above them or the water molecules below them.

The charged poles that pull water molecules together are the very feature that enables them to pull other molecules apart. Chemists speak of water as an unusually versatile solvent. Like the perfect dinner party host, water gently breaks apart couples (like sodium chloride) and large groups (like sugars and amino acids). Chemists also speak of water as a very good medium for diffusion. Again like that perfect host, water provides its guests an environment in which their parts can move and mingle freely. This environment, it should be said, happens to be particularly congenial to life. Water offers protection from DNA-damaging ultraviolet radiation, and it holds heat so well that temperatures near the ocean floor are unchanging year-round. And because, by comparison with other chemicals, water stays liquid at a very wide range of temperatures (in fact, a range of 100 degrees on the scale that is based on that very liquidity), life can operate at that same wide range.

One of water's properties is at once so peculiar and so conducive to life's presumed beginnings and long-term well-being that some nineteenth-century naturalists pointed to it as evidence of intelligent design.[17] If water were like most liquids, it would become denser and heavier when it froze. Ice would sink, and bodies of water in colder climes would radiate away heat and freeze solid from the bottom up. Life in those places—especially aquatic life—would have a very hard go of it. In fact, though, ice expands when it's frozen, becoming more voluminous by about 10 percent and forming a surface layer on lakes and oceans that insulates the water and organisms beneath.

As if all this congeniality weren't enough, water also uses dissolved compounds to make "microenvironments" within itself. The charged poles of water molecules lead other molecules to orient themselves side by side and facing in the same direction, some forming whole choreographed chorus lines, row after row of them, until they are best regarded as membranes. Some of these membranes develop into the microscopic bubbles that molecular chemists call *vesicles*, and whose interiors, some 4.6 billion years ago, may have sheltered the first self-replicating molecules and over time developed into cells.

Given all the virtues of water, we should not be surprised if organisms no one would call extreme go to astonishing lengths and employ ingenious strategies to get it. And they do. Spanish moss (*Tillandsia usneoides*) pulls water directly from the air; a species of kangaroo rat (*Dipodomys merriami*) draws it from metabolized food; and California redwood trees (*Sequoiadendron giganteum* and *Sequoia sempervirens*), by a means only imperfectly understood, pump it to their highest branches 100 meters above the forest floor. And once they have water, organisms no one would call extreme go to great lengths to hold it, to keep it from freezing or evaporating, to distribute it within themselves, and, where possible, to recycle it.

As for extremophiles? To acquire and retain water, they go to lengths that are, well, extreme.

FIRE AND ICE

The Celsius temperature scale uses the range at which water is liquid as its central scaffolding, but that range may be extended upward into hotter temperatures if, as we've seen, the water is kept under pressure. It may be extended downward into colder temperatures if the water is mixed with something else. Extremophiles are quite willing to exploit this wider range of temperature, and biologists are interested in the strategies they use to do it.

To appreciate the ingenuity of those strategies requires a brief refresher in biology. The *cell* is the smallest structural unit of an organism that can function independently. The cells in you and me, and in any other multicellular being, have a nucleus that contains their DNA. In the rather simpler cells of bacteria and archaea (groups to which all microbial extremophiles belong), the DNA floats freely in the semiliquid *cytoplasm*. In the cytoplasm are large molecules called *proteins* that initiate and accelerate chemical reactions in the cell and (in some cases) act as a supporting structure. The DNA, cytoplasm, and proteins are held inside a plasma membrane covered by a cell wall. The membrane protects what is inside the cell from the harsh environment outside it, and the wall prevents the cell, in certain situations, from expanding and bursting. The membrane and the proteins in the cytoplasm inside happen to be especially vulnerable to high temperatures. In water approaching boiling, the cell membrane grows more and more watery, eventually becoming too porous to do its job, while the proteins inside it are twisted, bent, or just plain broken (or as microbiologists put it, "denatured"), and so made useless.

To stay healthy in hot water, some thermophiles substitute the weaker parts of proteins with parts that are more durable and heat resistant. This is probably the method used by the hot-water record holder at present, a bacterium retrieved from a hydrothermal vent off Puget Sound. In 2003, University of Massachusetts microbiologists Derek Lovley and Kazem Kashefi had cultured the bacterium successfully and were curious about how much heat it could tolerate. They increased the temperature to 100°C, and the bacterium kept growing. The only means to still-higher temperatures that they had on hand was an autoclave, the pressurized steam–heated vessel commonly used to steril-ize medical equipment—an instrument, one can't help but note, designed not to culture microorganisms, but to kill them. They left the bacterium cooking in the autoclave for ten hours. The bacterium reproduced at 121°C and survived for two hours at 130°C. "We were," Lovley said, "truly amazed."[18]

There are reports of microbes, also living near hydrother-mal vents, that survive at still-higher temperatures, but collect-ing samples in the vicinity of hydrothermal vents is difficult, and the samples in question may have been contaminated. Still, since scientists can imagine substituting parts that would allow a cell to hold up under even higher temperatures, a confirmed find-ing would not be particularly surprising. As the NRC's *Limits of*

Organic Life report observes, "the upper temperature limit for life is yet to be determined."[19]

As to the lower temperature limit? Ice threatens an organism by an act of omission, denying the organism the solvent it needs to work its chemistry. It also threatens with an act of commission: ice crystals can easily tear a cell membrane. When water inside a cell freezes, the result is, in the ominous language of one paper, "almost invariably lethal."[20]

If water didn't mix well and life insisted on taking its drinks straight up, the coldest temperature at which an unprotected cell could survive would be 0°C, and we could learn all there was to learn about the chemistry of water in an afternoon. But as it happens, water will mix readily with any number of solutes. Stir in the right salts and you can keep water liquid at −30°C. Add some organic solvents and the temperature can go lower still. Where organisms can supply these salts and organic solvents, they will.* Some keep the juices flowing by increasing the concentration of solutes between cells; others, by modifying lipids and proteins in cell membranes. Mix in an amino acid like methionine and an organic compound like ethylene glycol and you can expect that enzymes, the proteins that act as biochemical *catalysts*, speeding up reaction rates, will still do their catalyzing at a chilly −100°C.

Ideas of how organisms might adapt to mixes of water and ammonia or water and liquid methane are what get biologists (and especially *astrobiologists*, who hypothesize about extraterrestrial life) excited about places like Saturn's moon Titan, where the warmest midday temperature might be −179°C, water

---

* It's worth noting that *organic* does not mean "living or once living"; it means "denoting or relating to chemical compounds containing carbon."

ice is hard as granite, and methane, a gas in our atmosphere, is cooled to a liquid. Exactly how low, under the limbo stick of temperature, can life go? The NRC report proffers that, given the right solvent, "it is possible that there is no low temperature limit for enzyme activity or cell growth."[21]

THE CHALLENGE OF SALT

It is not quite accurate to say that all extremophiles were identified after the discoveries of Corliss and Edmond. Some, including members of a group called "halophiles" (salt lovers), were found decades earlier. In the late 1930s, a graduate student named Benjamin Volcani, then studying at Hebrew University in Jerusalem, began to look for microorganisms in the Dead Sea. It was, to many, a curious pursuit. Hydrologically speaking, the Dead Sea is a closed-basin lake. In recent years, with the diversion of water from the Jordan River, its only substantial inflow, the Dead Sea has grown saltier and more alkaline. But even in the 1930s, its waters could be five times as salty as seawater, and often reached the point of saturation.[22]

The threat of salt water to a cell derives from the tendency of water molecules to balance the concentration of solutes on either side of a cell membrane. Salt water outside a cell will pull water from the cytoplasm inside through the membrane, and the cytoplasm will dry up.

In the 1930s, one would have had good reason to suppose that the water in the Dead Sea was lifeless, and many did. And so it came as no small surprise when Volcani found not merely a few organisms, but a thriving microbial community.[23] They had solved the salt problem, as do many archaea and bacteria in brine lakes everywhere, with an "if you can't beat 'em, join 'em" strategy, keeping high concentrations of salt in their cytoplasm, and so balancing the concentration inside against the concentration outside. But enough salt inside a cell can cause other problems. It will, for instance, bond with the water molecules that normally coat proteins, stripping them of that protection and making them vulnerable to denaturing. It turns out that the proteins in the cells of salt-loving archaea and bacteria have defenses—like, for instance, charged amino acids on their surfaces that hold on to the watery coating.

THE TEST OF ACID

On a shelf in her tidy, book-lined office in Woods Hole, biologist Linda Amaral Zettler keeps a small glass vial that she purchased in a tourist shop. It contains a few milliliters of what, in another setting, you might suppose to be red wine—perhaps a cabernet. In fact though, it is not quaffable, at least not by *mesophiles* like us. The liquid in the vial is a dilute acid laced with heavy metals, and it is from the Rio Tinto, a river in southwestern Spain.

The Rio Tinto's source is an iron ore deposit—or rather, what's left of one. The site has been mined, literally, since Paleolithic times, and what remains is a crater filled with water more acidic than vinegar. It is this acidity that dissolves iron, and it is the iron, oxidized by bacteria and exposed to air, that gives the water its reddish color—an indication of the high concentration of metals that the river maintains for all its 600-mile length, as it winds through rust-colored hills and scrub pines to empty, finally, into the Atlantic.

For years many had assumed the river was lifeless. As Dr. Amaral Zettler will tell you, they did not look very closely.[24] Even without a field microscope, anyone can see films of algae on seeping walls along the river's edges and, attached to rocks beneath the surface, green filaments of algae and whitish filaments of fungi waving in the current. But perhaps more surprising is what is living in and among the films and filaments. There are amoebas, ciliates, euglenoids, and flagellates—a thriving microbial community—not as diverse as that in a freshwater pond, but far more diverse than anyone expected.

Amaral Zettler is interested in many aspects of these organisms—one of which, quite naturally, is exactly how they manage to survive. Some set up defenses at the cell membrane, mostly with added proteins, that keep the inside at a more neutral pH and mostly free of metals. Others accumulate metals inside the cell, evidently without doing themselves serious harm. But research on the subject has barely begun, and it is probable—in fact, likely—that the microbial life in the Rio Tinto is protecting itself by other means as well.[25]

GOING WITHOUT

Readers of a certain age will recall an advertisement found in the back pages of many comic books, alongside the X-ray glasses and hovercraft plans, for "sea monkeys." An illustration promised an underwater city bustling with miniature creatures that looked a bit like chimpanzees, if chimpanzees had spiny dorsal fins and webbed fingers and toes. It was, so we readers were led to believe, a completely self-contained alien civilization we could keep on the dresser in our bedroom. What actually arrived in the mail was less miraculous, but only slightly. It was a small foil packet containing what looked like coarse-grained paprika. If you poured it into a glass of warm salt water and held a magnifying lens to the side of the glass, you would see tiny creatures uncurl, wriggle, and swim. In fact, they were brine shrimp (*Artemia salina*).

The shrimp had survived without water through a trick shared by many organisms—including bacteria, yeasts, fungi, plants, and insects. It is a process called "anhydrobiosis," by which cells shut down their whole metabolism and simply wait, as it were, for a rainy day. Some can wait a very long time. In the 1960s, archaeologists excavating Masada, the fortress in the Judean desert built around 35 BCE, found date seeds. Radiocarbon testing dated the seeds' shell fragments to the same period, and someone thought it might be interesting to see what would happen if the seeds were planted. Of the three, one germinated and soon grew into a healthy meter-tall plant.[26]

These remarkable examples notwithstanding, the undisputed champions of longevity are not any particular organism, but the dormant stage in the life cycles of many bacteria, plants, algae, and fungi. They are the small, lightweight, stripped-down versions of seeds known as *spores*. As a group, spores are profligate (a single mushroom may release millions), but as individuals they are downright spartan. They keep within them little, if any, stored food, and evidently they don't need much. In 1995, scientists resuscitated *Bacillus* spores that had been trapped in amber at least 25 million years.[27] And spores are inventive, making salt (a cell's enemy) into a shield. When salt water evaporates, it may leave deposits that have, trapped within them, tiny pockets of water called "brine inclusions," microenvironments in which spores can survive. A *Bacillus* spore has been reported revived from brine inclusions thought to be 250 million years old—older, that is, than the first mammals.[28]

To many scientists in the late nineteenth century, spores seemed overengineered, far tougher than they needed to be; and some wondered whether they might have evolved in an environment much harsher than any on Earth. If they did, then they might explain life's origin.

In the early nineteenth century, many natural philosophers held that organisms arise by spontaneous generation from organic matter. In 1860, French chemist and microbiologist Louis Pasteur conducted a series of experiments involving much care and many flasks and filters, and demonstrated that such could not be the case. Two possibilities for life's origin on Earth remained: either life had arisen in the distant past, in the form of organisms far simpler than any in existence in 1860; or it had come from somewhere else. The second hypothesis, now termed *panspermia*, was put forth a few years after Pasteur's work by Lord William Thomson Kelvin, who suggested that life originating on another world may have arrived on Earth via "seed-bearing meteoric stones."[29]

Such a trip would not be easy. Suppose it were from Mars to Earth. An organism, actively metabolizing or dormant inside a fragment of Martian rock, would have to be well positioned—not so near a meteor strike that it would be vaporized, but near enough that it could ride the blast's shock waves (and withstand tremendous g-forces and heat) up through the atmosphere and into interplanetary space. Once in space, it would have to survive vacuum, radiation, and extremes of temperature, and it would have to do so for years, decades, or perhaps centuries. Finally, it would have to withstand a fiery entry, along with more g-forces, into Earth's atmosphere, ending its journey with an arrival violent enough to leave a crater.

Given spores' well-known feats of endurance, many astrobiologists have wondered whether they might be up for the trip, and a few have devised experiments to simulate one. If you were a spore, you might regard astrobiologists as the sum of all fears. Astrobiologists have baked spores, frozen them, irradiated them, fired them from guns, and slammed them between quartz plates with explosives. And in case such simulations fell short of the rigors of actual space travel, they placed them aboard NASA's orbiting Long Duration Exposure Facility and left them outside the spacecraft, unprotected except for a thin aluminum cover, for six years. At present, despite its advocacy by several respected scientists, panspermia lacks widespread support. Nonetheless, the upshot of all these experiments is that spores can withstand a violent launch and reentry, and that as long as they are shielded from ultraviolet radiation with a few centimeters of soil or rock, they are quite capable of surviving in space for decades—long enough for travel among planets within the Solar System. If life on Earth did come from elsewhere, it could have made the journey as a spore.

NOTHING LIKE THE SUN

Microbes collectively called "intraterrestrials" have been found several kilometers beneath Earth's surface, making for a kind of subterranean *biosphere*.[30] Bacteria have been found in rock samples taken several hundred meters below the seafloor, even in places where the seafloor itself is several kilometers below sea level. No one knows how many organisms are living in this environment, but the number may be large. One recent study found between a million and a billion bacteria per gram of rock. It may be that a large proportion of all bacteria on Earth live below the floor of the sea, where their metabolisms are driven by energy from various sources (like natural radioactivity) that are utterly independent of the Sun. But even extremophiles on Earth's surface have been discovered exploiting unusual energy sources. One fungus was found in the water core of the Chernobyl nuclear reactor, ingeniously and fearlessly converting nuclear radiation into usable energy and managing radiation damage by keeping copies of the same chromosome in every cell.[31]

THE PRESENT

A list of extremophilic world record holders that elicits a "wow" also risks a dismissal—an assumption that they are freaks in a biological sideshow, having little to do with biologists' larger interests. In fact, though, there are real, baseline reasons to count extremophiles as important players in the epic of life on Earth. Until the late twentieth century, many biologists supposed that all life on Earth began in the "warm little pond" that Darwin's contemporaries favored or its more sophisticated successor, the *"prebiotic* soup" that Stanley Miller and Harold Urey tried to replicate in their famous experiments in the 1950s.[*] These suppositions and many others, along with a century or so of thought, were challenged when, three years after Corliss and Edmond discovered hydrothermal vent communities, Corliss and a group of colleagues published a paper arguing that life might have begun in or near a hydrothermal vent.[32] Recent evidence suggests that thermophiles much like those now living near the vents may have been the ancestors of all life on Earth.[33]

These discoveries come at a time when many mesophiles are being discovered and catalogued for the first time. Especially in a moment when species are being made extinct at a terrifying

---

[*] Experiments enshrined in textbooks but whose presuppositions about the chemistry of the early Earth's atmosphere are now largely discredited.

rate—exceeding that of the five great extinctions in the last half-billion years, and at least a hundred times faster than the normal background rate—it may come as a surprise to learn that since 2005, about 400 species of mammals have been newly identified. But this is not necessarily good news. Many were discovered precisely because, with their habitats destroyed by logging, human settlement, climate change, pesticides, invasive species, and so on, they were disturbed, made suddenly visible—and vulnerable. We are, as it were, burning down the forest and watching to see who runs out.

It is here that extremophiles bring cause for a kind of big-picture optimism. If individual organisms and whole species are fragile, then life in general is resilient, tenacious, and, in its willingness to exploit any and all environments, downright aggressive. It is also inventive. When a suitable environment does not exist, life may create one. The most extreme extremophiles are of the domain called Archaea—the domain whose members were the first life on Earth, and a billion or so years from now, when our ever-warming Sun will have baked the ground and boiled away oceans, are likely to be the last. Even now, if the worst happened and a nearby star exploded, roasting Earth with gamma rays and exterminating all life on the surface and in the upper layers of the oceans, those assemblages of bacteria and fungi living a kilometer deep would go on as if nothing had happened. In time they would colonize the surface, probably learn the trick of synthesizing sunlight, and start things all over again.

Certainly the resulting scenery wouldn't satisfy the aesthetic of, say, nineteenth-century American landscape painters. But then again, the assemblages of bacteria and fungi probably wouldn't much care for the aesthetic of nineteenth-century American landscape painters—or ours, for that matter. And yet they and we are distant relatives. In fact, all life we know shares certain basic features. If you could take a cell from any organism—an alga, a giant sequoia, a condor, or your second cousin—and dive through its cell membrane and into its cytoplasm, you would find precisely the same nucleic acids and proteins doing precisely the same things in precisely the same ways.[34]

In fact, these shared features are what lead evolutionary biologists to suspect that everything that lives and has ever lived is descended from a single common ancestor, a microbe that metabolized some 3.5–3.8 billion years ago and (luckily for us) reproduced.[35] You might expect that, given its role as the very origin of life on Earth, this microbe would have been granted a name evocative of grandeur and myth. But, perhaps doing an end run around cultural politics, or realizing that any name would have to be borrowed from one of the microbe's descendants, biologists call it, somewhat prosaically, the *last universal common ancestor*, or *LUCA*.[36]

What interests a great many biologists is that many of the features shared by all known life seem to have no "selective advantage." In other words, it didn't have to be this way. There were, and are, alternatives. Chemists can imagine billions of organic compounds, but life uses only about 1,500. Those working in the new field of *synthetic biology* can imagine other amino acids, other proteins, and other metabolisms (or at least parts of metabolisms) that use other processes and would work just as well, perhaps better.

Quite naturally, a question arises. Was LUCA truly universal? Might there have been, in the 4.6 billion years of Earth's history, a second genesis—a moment when complex molecules gave rise to another living organism, independent of and unrelated to LUCA? Might this organism have then reproduced? Might it even have established a line of descent that has endured as microscopic, single-celled Sasquatches into our own time? And if such organisms exist, since they arose from a chemistry different from that which produced LUCA, might they survive and even flourish beyond the limits of the most extreme of extremophiles?

These are profound and haunting questions, and they much occupy the thoughts of the several scientists (and one philosopher) we will meet in the next chapter.

CHAPTER TWO

A Shadow Biosphere

D arwin was excruciatingly careful to distinguish what he knew from what he did not know, and to distinguish both from what, given the limits of biology in his day, he *could* not know. In an 1871 letter to botanist Joseph Hooker, Darwin refers to the then fashionable idea of an origin for life in "some warm little pond." But contra many who have taken the phrase out of context, Darwin did not claim the idea as his own, and he observed later in the same letter, "It is mere rubbish thinking at present of the origin of life; one might as well think of the origin of matter."[1]

If anything, the origin of life has proved the more difficult problem. Since the mid-1920s, biologists have agreed that life is the product of complex chemistry, but other aspects of the subject have been, and continue to be, vigorously debated. Many have conjectured as to its place of origin: that "warm little pond" and variations like the ocean and drying lagoons, surfaces of clays, deep-ocean hydrothermal vents, mineral surfaces of ice veins in glaciers, the pores of rocks deep within the Earth, even clouds. As mentioned in the previous chapter, some have suggested that life began elsewhere in the Solar System and was delivered to Earth via meteorite. There have been at least as many ideas as to its first form: enzymes, viruses, genes, and cells, to name a few. All these ideas have played out against the more fundamental question of life's sheer probability, with the pendulum of informed opinion swinging on a wide arc between "improbable in the extreme" and "almost inevitable." About the only point on which there has been general agreement is that if we could trace the ancestry of all living organisms back far enough, we would find them converging, some 3.5–3.8 billion years ago, at a single genesis. Life on Earth, so most scientists believe, began at one place and at one time.

Most scientists, but by no means all. Some suspect otherwise, and their reasoning is quite straightforward. Since, as the vast majority of biologists now believe, life is not a once-in-the-history-of-the-universe event, but a more or less inevitable by-product of physics and chemistry, it follows naturally enough that life on Earth may well have had more than one beginning. It follows further that if a second beginning had occurred under even slightly different circumstances, a different sort of life would have resulted.

This is the possibility described and explored in a 2009 article called "Signatures of a Shadow Biosphere."[2] Its six authors represent, as we might expect, a rather unusual collection of expertise. Four of the six have backgrounds in the life sciences. Two of them—perhaps the two who have worked hardest to bring the article's provocative ideas to a wider audience—are cut from a rather different disciplinary cloth. Paul Davies trained as a mathematical physicist, and as recently as the 1990s his main work was in cosmology and quantum gravity. Of late he has widened his gaze considerably, becoming more interested in fundamental questions about the nature of scientific inquiry. Carol Cleland is a member of NASA's Astrobiology Institute and—this belying any charges that NASA lacks a wider perspective—a professor of philosophy at the University of Colorado, Boulder. She is fond of quoting Thomas Kuhn (the historian of science we met in the previous chapter) and suspects that many scientists miss opportunities because they don't think outside the box Kuhn calls the "reigning paradigm." It is Cleland, along with microbiologist Shelley Copely, who coined the phrase *shadow biosphere*—a provocative and slightly unsettling reference to a hypothetical biosphere of microbial weird life that, like the realm of fairies and elves just beyond the hedgerow, may or may not impinge on our own.

Davies, Cleland, and the other authors of the article are excited by this possibility for two reasons. First, the discovery of such life would make it possible for biologists, by comparing the differences and commonalities of two examples of life, to begin to discover universal laws of biology much as physicists since Newton have discovered universal laws of physics. As a science, biology would have fully matured. Second, and more profoundly, the discovery of such life would settle the debate over life's probability once and for all. It would mean that life in the universe is common and may arise anywhere conditions are right. Not incidentally, such a discovery would ripple far beyond biology into all realms of human experience, altering forever our understanding of our place in the universe.

But we're getting ahead of ourselves. If a shadow biosphere of weird life exists, it would be prudent, before confronting questions of its larger meaning, to ask where and how it might have begun.

A SECOND GENESIS

The scarred and battered face of Earth's Moon is the visible legacy of the violence of the early Solar System, a time some 4 billion years ago when asteroids and comets routinely struck all its larger worlds, including a molten, slowly cooling Earth. Heat and radioactivity from the planet's core sent lava through cracks in the newly formed crust, and as the surface cooled, steam from the atmosphere condensed and fell as rains that, lasting for millennia, created the planet's first, shallow oceans. Such an environment would seem inhospitable to life, yet the most necessary ingredients were there: complex carbon-based molecules and liquid water. Indeed, most scientists believe that this is the environment in which life gained its first foothold.

Almost all familiar life builds its proteins from the same twenty *amino acids*—the molecules that biology textbooks call life's "building blocks."* What's interesting is that an organism would enjoy no particular advantage by limiting itself to these twenty, and many others might work as well. It seems that in its first incarnations, familiar life used the amino acids it used for no better reason than that they were available and nearby. In another part of the early Earth (and Earth, it is worth remembering, is a big place—still bigger if you're a few complex organic molecules edging toward replication and self-organization), other amino acids would be available. And another set of complex organic molecules edging toward replication and self-

---

* The equivocation in that word "almost" derives from the fact that scientists actually know of twenty-*two* naturally occurring amino acids on Earth. The genetic code of certain organisms can include selenocysteine and pyrrolysine, although the latter has been found so far in only one organism—an archaean called *Methanosarcina barkeri*.

organization might use them—the result being a second genesis, of another sort of life.

In 1988, Caltech geologists Kevin Maher and David Stevenson suggested that the standard picture of life's beginning was too simple.[3] Their point was that conditions suitable for life may have lasted many millions of years; there was world enough and time for many beginnings—but perhaps no more than beginnings. The reason for that last qualification is that the era in Earth's history when life began overlapped with periods of "heavy bombardment" by meteors. Every so often—on average once in a half-million years—an unusually large meteorite—say, the size of Manhattan Island—would strike with such force that the oceans would boil, the atmosphere would be superheated, and the planet would be rendered all but sterile. The time between each of these armageddons might be just enough for life to begin all over again. But if one beginning during any particular respite is unlikely, two beginnings are improbable in the extreme. And for this reason we might conclude that there is little chance that at any given time two forms of life coexisted. We might conclude this, that is, except for the fact that no particular sterilization would be complete. After all, familiar life today survives and flourishes on the ocean floor and deep underground—both places well protected from any unpleasantness nearer the Earth's surface. A robust sort of primitive organism might have done likewise.

Sheltered locales on and in Earth are not the only places such an organism might have weathered the storm. Davies has suggested another, rather more distant, refuge. A meteor striking Earth with sufficient force might launch fragments of rock into orbit around the Sun. Some meteors might hold microbes or spores that could lie dormant for thousands or even millions of years, until the moment when the orbit of the rock fragments and the orbit of Earth happened to intersect, and the fragments would fall back to Earth. Some would split open on impact, and their microbial passengers—any that survived, that is—would wake to a world once again fit for life, and perhaps already harboring another kind of life, one that had appeared in the half-million years during which they were away. Like Homer's Odysseus, the microbes would have returned home after a voyage of many years, to find strangers living there. It would be a space odyssey on a microbial scale—and an interplanetary one.

As mentioned earlier, life on Earth might not only have returned from somewhere else. It might have *begun* somewhere else. Four billion years ago, the planet Mars had a thick atmosphere of carbon dioxide, with rainfall, and streams and rivers of liquid water coursing through valleys and emptying into lakes and shallow seas. In short, it was a congenial abode for life. Like Earth of the period, Mars was also pummeled with meteors, and some struck with enough force to launch rock fragments into orbit around the Sun. After thousands or millions of years, some of the fragments intersected Earth's orbit and fell to Earth. In fact, scientists have found at least twenty-eight of them, one of which is ALH84001, made famous in 1996 when David McKay, chief scientist for astrobiology at NASA's Johnson Space Center, and his research group suggested that it bore evidence of life. Although their conclusions remain controversial, it seems clear that early in their history, Earth and Mars traded material, and some of that material may have contained microbes. Davies and others believe it barely possible that life on Earth—familiar, weird, or both—has a Martian ancestry.

None of these ideas are proof that weird life, let alone a shadow biosphere of weird life, exists. But collectively, they make a case that weird life had ample time to arise on Earth and several means by which to do it. Suppose then, that it did arise. An obvious question presents itself: Wouldn't we have noticed by now? And if it were microbial, wouldn't microbiologists have noticed? The answer, interestingly enough, is: not necessarily.

WHAT WE DON'T KNOW

Those of us who take our science news from *Discover* magazine and nature shows are regularly and properly astounded by what biologists and microbiologists know. If we were to learn what they *don't* know, we might be just as astounded, for what they don't know is a great deal. Take, for example, the answer to the straightforward question, How many species are there? The difficulty here is simply that there is no reliable way to determine that number, or even to estimate it, except perhaps to replicate work performed in 1981 by Terry Erwin of the Smithsonian Institution.

Erwin wanted a census of the number of the world's arthropod species—insects, spiders, crustaceans, centipedes, and the like. He and his team arranged a grid of specimen bottles, with 1-meter-wide funnels affixed to each, beneath a tree in Panama. With the air calm, they sprayed insecticide into the canopy, and some hours later they collected and began to classify the thousands of arthropods that had fallen through the funnels and into the bottles. Erwin counted 163 species of beetles known to live exclusively in the species of tree they had fallen out of, multiplied that number by the number of tropical tree species known, and concluded that beetle species numbered more than 8 million (thus incidentally supplying quantitative evidence for British geneticist J. B. S. Haldane's possibly apocryphal remark that the Creator must have "an inordinate fondness for beetles"[4]). Since beetles are known to represent 40 percent of all arthropods, Erwin assumed the same proportion in the tree whose denizens were under study, and after a number of other calculated guesses, he concluded that the number of arthropod species worldwide might be as high as 30 million.[5] But no one, including Erwin, thinks this number definitive, and other estimates vary wildly.

Bear in mind, too, that this is only what we don't know about just one phylum. Our ignorance of the rest of the natural world is proportionately greater. In 2002 the famed entomologist Edward O. Wilson estimated that 1.5–1.8 million species have been identified and catalogued, but well-reasoned guesses of the actual number lay within a stunningly wide range: 3.6–100 million.[6] The full meaning of these numbers is so dumbfounding as to bear restating: for every species known to science, there is at least one that is unknown, and there may be as many as fifty.

Since Erwin's work, several international programs have begun to catalogue biodiversity. The Census of Marine Life, a decade-long project to make a comprehensive tally of life in Earth's oceans, found 5,000 previously unknown species, including an animal that lives without oxygen, several species believed to be extinct since the Jurassic period, and 600-year-old tube worms. The ongoing International Barcode of Life project identifies species with only a snippet of their DNA, and has so far assigned bar codes to more than 100,000 species. Coordinated with both projects and with several zoological organizations is the *Encyclopedia of Life* (mentioned earlier), now at half a million pages and growing.

Of species still undiscovered, it is possible that some are quite large. As recently as the mid-1990s, scientists were astonished to discover a 200-pound animal inhabiting the mountains shared by Vietnam and Laos. It looks part antelope and part cow but, now classified as the only member of the genus *Pseudoryx*, is neither. Most unknown species, though, are likely to be small—and many are no doubt microscopic. The 1989 edition of Bergey's *Manual of Systematic Bacteriology* lists roughly 4,000 species of bacteria,[7] but microbiologists, using several ingenious and indirect measurements, have inferred that the true number may be in the millions.

Our ignorance of the microbial realm is disquieting—or should be—not merely because there is so much of it (microbes compose as much as 80 percent of the Earth's biomass and 10 percent of your dry body weight), but because it is the realm from which our own "macrobial" realm originated and upon which it still depends. Microorganisms act as the basis of all food chains and work to regulate the chemistry of Earth's atmosphere and oceans. In fact, if the *Gaia hypothesis* of British inventor and scientist James Lovelock and American biologist Lynn Margulis has any validity, then Earth's climate has for billions of years been held in delicate equilibrium by oceanic phytoplankton and other microorganisms working, one must note, without committees, treaties, or international protocols. Their other achievements are similarly impressive. They originated all the chemical systems upon which life depends, systems we cannot yet replicate and do not fully understand. And they have adapted to the most extreme of Earth's environments—environments in which we, without artificial means anyway, could not survive. Microbes were the first organisms on Earth, and given their record of success, they will surely be the last.

The reason we have so little knowledge of the microbial world lies in the limitations of the instruments and techniques we have available to explore it. Under a microscope, that most time-honored of scientific tools, a given species from the domain

Archaea and a given species from the domain Bacteria may be indistinguishable, even though they have less in common with each other than you have with, say, a soft-shell crab. Most bacteria and archaea look like spheres or rods. Microbiologists can enhance the view and identify parts of any given cell with "staining," but the parts might represent only some differences, and not necessarily the most important or fundamental ones.

Microbiologists who want to study a microbe thoroughly and over time will "culture" it—that is, introduce a sample of the microbe to nutrients in standard culture dishes, and wait until the sample proliferates into a colony containing enough individual microbes that they can be sorted and analyzed. This is not as easy as you might expect. While certain species, most famously *Escherichia coli*, grow so readily that laboratory biologists call them "weeds," the fact is that most single-celled organisms don't survive long in captivity. Many a microbe that thrives in a puddle or pond, when carefully removed, carefully transported, and carefully placed in a culture dish, will shrivel up and die. To a nonscientist, it may come as a shock to learn that biologists have been able to culture less than 1 percent of the microbes they have seen, as it were, in the wild.[8]

Not that they know all that much about the wild. With humility that one can only call admirable, the NRC report of 2007 notes, "It is clear that little or nothing is known of the physiological diversity of most microorganisms in most Earth environments."[9] This includes environments that are nearby. As Wilson observes, a pinch of soil from any forest floor, no more than can be held between thumb and forefinger, is likely to contain thousands of bacterial species, many of them unknown.[10]

All this is by way of saying that the fact that we have not found microbial weird life should not lead us to conclude that it doesn't exist. As English Astronomer Royal Martin Rees observed, with regard to another scientific mystery, "Absence of evidence is not evidence of absence."[11] Or is it?

I add a dollop of doubt because we might easily imagine a second objection to the notion of weird life. Familiar life is successful, as mentioned earlier, because it is resilient, tenacious, aggressive, and inventive. Suppose that at some moment in the roughly 4-billion-year reign of familiar life, a sort of weird life *did* emerge. Might we assume that it would have lost any and all competition for resources, and that almost immediately after its appearance, familiar life would have pushed it into extinction? The answer, again, is: not necessarily. According to Davies, Cleland, and their colleagues, there are at least three ways weird life might have managed, and might manage still: as ecologically separate from familiar life, ecologically integrated with it, or biochemically integrated with it.

THREE TYPES OF SHADOW BIOSPHERE

One way weird life could manage is by moving into places that no familiar life, not even extremophiles, wants. There are many such places—the core of Chile's Atacama Desert,[12] ice sheet plateaus, hydrothermal vents with temperatures above 400°C, and high-brine liquid water at temperatures below −30°C. Weird life in any of these places would likely be part of a biosphere *ecologically separate* from our own—and these are phenomena known to exist. Since 1990, scientists have discovered several ecosystems of extreme familiar life that are separated from the rest of the biosphere. There is a microbial community beneath the Columbia River in Washington State composed of bacteria that live inside basalt rock, another in the Twin Falls area of Idaho, still another near a gold mine in South Africa.[13] Each is remarkable for its source of energy: chemosynthesis in the first two cases, and radioactive decay in the third.

There is also the possibility that weird microbes, while greatly outnumbered by familiar microbes, are living among them. Molecular biologist Mitch Sogin, a coauthor of the 2007 NRC report, called the diversity of most microbial communities "staggering," and noted that most of the diversity was owed to a small number of individual microbes.[14] In other words, few microbes of each species, but an enormous number of species nonetheless. It is possible that weird life is present and unaccounted for in many microbial communities, keeping its profile low and, since it is weird, consuming what no one else wants and excreting what no one else is bothered by. Such weird life would compose a biosphere *ecologically integrated* with our own.

Finally, there is the possibility—this perhaps the strangest of all—of weird microbes and familiar microbes in *symbiotic* relations that benefit both, trading chemical compounds, enzymes, or even genes. Symbiotic relations in the microbial realm have a long history—a history demonstrating that, contra ideas of nature as "red of tooth and claw," there is as much cooperation as competition, and perhaps a good deal more.* Consider the strange case of mitochondria, the *organelles* that perform respiration and generate chemical energy. It is thought that some 3 billion years ago they were oxygen-respiring purple bacteria and microbial nomads, finding comfort and sustenance where they

---

* The idea of cooperation between species was (of course) not lost on Darwin, who noted, "A flower and a bee might slowly become, either simultaneously or one after the other, modified and adapted in the most perfect manner to each other, by the continued preservation of individuals presenting mutual and slightly favourable deviations of structure." (*Origin of Species*, 85)

could, and otherwise making do in a harsh world. Then, one or several of them found refuge in the warm, wet, pH-balanced interior of a cell, and took up permanent residence. Others followed, and achieving survival more by snuggle than struggle, host and guest eventually negotiated terms. The cell provided the bacteria protection, and the bacteria supplied the cell with oxygen-derived energy and disposed of its waste. In the fullness of time, the arrangement developed into a codependency so complete that today, the cells in your body would die without the mitochondria inside them.

If weird microbes exist, it is possible they've established similar arrangements with familiar microbes. They would comprise a biosphere *biochemically integrated* with our own. If ecologically separate weird life is the person you'll never meet, and ecologically integrated weird life is that utterly silent and all but invisible boarder, then biochemically integrated weird life is the roommate who shares your toothbrush, borrows a twenty from your wallet and forgets she did it, but at regular intervals thoughtfully leaves a bouquet of flowers and a bottle of wine on the kitchen table.

At least in theory, there is no good reason to suppose that weird life doesn't exist on Earth. Suppose, then, that it does. The prospect is exciting for all the same reasons that the prospect of life on other worlds is exciting—perhaps more so, for the simple reason that weird life on Earth might be easier to find.

The search for life on other worlds—which began in earnest with recommendations from NASA subcommittees in the 1960s—has proved more challenging than many had anticipated, and it is unlikely to yield results anytime soon. The difficulty lies in the distance between researchers and their possible subject. Earthbound astronomers using long-range detection techniques like spectrometry can examine the atmospheres of planets and moons in our Solar System—and when conditions are right, some planets in other systems—for chemical compounds commonly called "biosignatures" that may have been produced by living organisms. But without on-site study by astronaut-scientists, sample-return missions, or at the very least, sophisticated unmanned probes, they can't know whether such compounds are true biosignatures or merely the product of an exotic chemistry.* To date, the only *in situ* search for life elsewhere came in NASA's *Viking* missions—with results that were inconclusive. The next-generation Mars Science Laboratory, which began its journey to the Red Planet in late 2011, is designed to answer questions about how well the Martian environment is suited to life, not to seek life directly. At the time of this writing, missions to Mars and elsewhere designed specifically to look for weird life are distant prospects at best.

By way of contrast—and this is a point Davies and Cleland make rather tirelessly—a systematic search for weird life on Earth could begin immediately and at a far lower cost. The only real question is how best to go about it.

SEEKING WEIRD LIFE ON EARTH

In a search for weird life on Earth, the standard tools and techniques for identifying microbes are unlikely to be of much help. The similar appearance of archaea and bacteria under a microscope suggests that their shapes—spheres and rods—have real evolutionary advantages, and we can expect that weird microbes

---

* The same uncertainty surrounds the detection of trace amounts of methane in Mars's atmosphere, which may indicate life but may also be produced by a geochemical process. (Tenenbaum, "Making Sense of Mars Methane")

will look much the same. Staining can highlight gross features of cells, but it can miss smaller ones, and these might be the very features that make the cells weird. Attempts to culture weird microbes would be especially challenging. Microbiologists trying to culture familiar microbes must make an educated guess as to the microbes' needs in the way of temperature, humidity, and nutrients. As to the needs of weird life, they might have no idea. It is true that there is a relatively new tool used to identify microbes, called "DNA amplification." But it works only if the DNA in question uses the sugars and bases of familiar life. It also works, of course, only with a microbe that has already been isolated. It would be of little use in distinguishing a weird-life microbe from the thousands of species of familiar life in that pinch of soil from the forest floor.

For that, Davies proposes a general rule of thumb: the more fundamental an organism's differences from familiar life, the greater its chances of being weird. For instance, if an organism uses a different amino acid, it is probably an unusual form of familiar life. But if it uses ammonia (not water) as a solvent, or silicon (not carbon) as a binding molecule, it is almost certainly weird. The hard calls would be in the middle, and one reason to expect some in the middle is a phenomenon called *convergent evolution*. This is the process by which two species respond to the same environmental challenge and take advantage of the same environmental circumstance by developing features that are similar—and in some cases identical.

The eyes of humans and octopi are an oft-cited but nonetheless remarkable example. Even in their details the two eyes are astonishingly similar, yet the fact that one sort belongs to a cephalopod mollusk with eight sucker-bearing arms, a saclike body, and a beak and the other sort belongs to a species of primate means that they evolved along entirely different evolutionary lines. Those lines converged because the need to detect predators and prey at a distance is well met by a feature sensitive to electromagnetic radiation in the visible spectrum. In fact, the advantages of sight are so pronounced that eyes evolved independently in marine worms, mollusks, insects, and vertebrates—organisms whose common ancestor was sightless. Convergent evolution, then, is a powerful force, and it is known to operate at the cellular level. Some enzymes in familiar life are remarkably similar, yet have entirely different ancestries. If convergent evolution operates for weird life (and there is no obvious reason it should not), then forms of weird life and forms of familiar life, while radically different from each other when they first appeared, may have grown so alike over time as to be nearly indistinguishable.

A scientist verifying an organism as weird faces yet another challenge—this having to do with the nature of life's beginnings. Some biologists suspect that the transition from nonliving to living (that is, from complex chemistry to simple biology) was abrupt, akin to the phase transition of water as its temperature is lowered through the freezing point and it crystallizes—the moment at which its molecules suddenly snap to attention in rigid lattices. If one could define life as, for instance, having the ability to store and process information, one could establish a similar boundary. On one side would be complex chemistry that could not store and process information; on the other would be simple biology that could. The transition from one to the other, had anyone been around to witness it, would have been unmistakable. And if it happened a second time, even with slightly different results, it would have been just as unmistakable.

A well-defined transition would mean that scientists who discovered a candidate for weird life might trace its lineage to the moment of transition with some hope of success. But if, as others suspect, the transition was gradual—a long series of steps, some quite small, and no particular step of which anyone could say with certainty, "This is where chemistry ends and biology begins"—then scientists tracing the lineage of weird life would have no hope of identifying a point and moment of origin. Of course, neither would they have any hope of identifying a point and moment of familiar life's origin. To follow either line would be like following two rivers upstream and finding that both began in a single network of smaller streams and rivulets, and that these were fed in turn by moving groundwater. It would be impossible to identify precisely where either river began, and it would be impossible to say whether they arose from separate sources. In fact, it would be pointless even to try.

Again, we may be getting ahead of ourselves. Before we trace an organism's provenance and make a case for classifying it as weird, we need to find it. How then to begin? Davies and his colleagues recommend designing searches targeted around a particular type of shadow biosphere. If, for instance, we're looking for weird life in a shadow biosphere that is ecologically separate, we might look for that separation. Suppose we discover a community of extremophiles in 200°C water ringing a hydrothermal vent. If we found that the hotter water just inside the ring and nearer the vent was sterile, we might reason that the inside edge of the ring marks the upper temperature limit for these particular extremophiles. But suppose that even nearer the vent, where the water is hotter still, we found, after minding the gap, a second ring of living organisms, clearly separated from the first. We would have some distance to go to prove it, but we would have reason to suspect that life in the second ring was weird.

If on-site identification proves difficult—and in these locales it often is—then Davies and his colleagues suggest we retrieve a sample of water, soil, or ice from a place too harsh even for extremophiles and, difficult as the prospect might be, try to culture any microbes present and wait for signs of life. Exactly what signs of life?

Steven Benner, another coauthor of the NRC report, has some ideas. Benner is a fellow at the Foundation for Applied Molecular Evolution, an organization whose rather audacious name is likely to prompt a few late-night discussions: Can we really apply evolution? Should we? Whatever the answers, the startling fact is that in the last twenty-five years, Benner and his colleagues have engineered several artificial biological components and systems. They have, for instance, synthesized a gene for an enzyme and built proteins with amino acids not used by natural proteins. Their work has practical benefits, having led, for example, to improvements in medical care for HIV patients. It might also be used in somewhat more arcane pursuits, like guiding searches for weird life. This because not only can Benner and his colleagues identify the parts of an organism that might be vulnerable to extreme conditions; they can also imagine substitutions for those parts. And because nature has had at least a 3.5-billion-year head start, so the thinking goes, anything Benner and company can imagine might already be out there somewhere.

For instance, the upper temperature limit for some hyperthermophiles is set by some of their amino acids, which denature at higher temperatures. Benner knows of another amino acid—2-methylamine acid—that folds in such a way that it can withstand those temperatures. If you are seeking weird life, you might retrieve a water sample from a place too hot even for hyperthermophiles, take it into the lab, and test for 2-methylamine acid. If you find it, you may also find weird life.

Alternatively, you might look for substitutions in the parts of DNA. Recall that if the DNA molecule is a flexible ladder whose ends have been given a few twists, then its long backbone (the two legs of the ladder) is made of sugar and phosphate molecules, and its rungs—all 3 billion of them—are made of chemicals called bases. There are four, and when the DNA molecule is intact, each is paired to its complement: adenine always with thymine, and guanine always with cytosine. This much is taught in any introduction to biology. What is seldom taught—and what might be of interest to seekers of a certain sort of weird life—is that the bases are what limit the pH levels tolerable for many extremophiles. Acidophiles can stand only so much acidity because the bases adenine and cytosine are relatively alkaline, and alkaliphiles can tolerate only so much alkalinity because thymine and guanine are relatively acidic. If weird-life DNA used different bases, it could withstand pH levels more extreme than those tolerated by known extremophiles.

ARSENIC

The weird life of an ecologically separate shadow biosphere might differ from familiar life in another fundamental way: its chemical composition. The fact that our bodies and the bodies of all life we know are made of a few simple chemical elements has been much used as a hard lesson in humility, a "to dust ye shall return" for secular types. But perhaps the better lesson is that the whole can be greater, much greater, than the sum of the parts. The whole—here meaning incredibly complex structures like proteins and lipids—is ordered almost entirely from a spare menu of six chemical elements: carbon, hydrogen, nitrogen, oxygen, sulfur, and phosphorus.*

"Phosphorus" means "light-bearing," and although we in the macroscopic world know it to be capable of fireworks, in the living cell it stores and transfers energy slowly and (one might say) carefully, as part of the chemical compound adenosine triphosphate, or ATP. It has other roles too, most notably in the phosphate (a molecule of one phosphorus atom and four oxygen atoms) that, along with sugar molecules, goes to make the spiraling backbone of the DNA molecule. What is interesting to weird-life research is that the roles of phosphorus could be performed as well by an element with a rather more sinister reputation: arsenic.

Arsenic is notorious as a poison and, perhaps as befits its part in many a murder mystery, it works on a biochemical level by stealth, mimicking phosphorus so well that it can gain entrance to a cell and make its way into metabolic pathways. Once inside, it turns ATP's careful distribution of energy into exchanges that are more explosive—and destructive. Nonetheless, like phosphorus, arsenic can bond molecules and store energy. If, some billion years ago, a set of complex, self-organizing prebiotic molecules was in need of an ingredient to do what phosphorus does for familiar life, and it happened to be in a place where phosphorus was rare but arsenic was plentiful, it might well have used arsenic for all its bonding and energy-storing needs—assuming,

---

* There are also trace elements, like iron and zinc, for which many organisms will make substitutions. Some mollusks, for instance, carry oxygen in their blood not with iron (the standard choice), but with copper.

of course, that it could develop means to cope with arsenic's instability.

It is worth noting that an organism using arsenic in the roles that familiar life gives to phosphorus would regard phosphorus as poisonous. If life had taken a different course, then we—or weird-life versions of us—might be suffering through summer stock productions of *Phosphorus and Old Lace*. But even given the course we know familiar life to have taken, it is possible that a second genesis of life chose arsenic, or that an early offshoot of familiar life substituted arsenic for phosphorus. It is also possible that in hydrothermal vents, hot springs, and closed-basin lakes— all places poor in phosphorus but rich in arsenic—it might still be hanging on.

In fact, this was the hypothesis that, in 2007, was put forth by a young postdoctoral researcher named Felisa Wolfe-Simon. She was already something of an iconoclast, having begun a career as a musician (trained as an oboist) but in time having earned a PhD from Rutgers in oceanography. In 2007 she was present in a workshop on weird life convened at Arizona State University by Davies, who was newly arrived there and laying groundwork for a research center that would address fundamental questions in science. Davies recalled, "We were kicking vague ideas around, but she had a very specific proposal and then went out and executed it."[15]

Wolfe-Simon's proposal had to do with Mono Lake, a closed basin in California's high desert some 20 kilometers across. Waters from the Sierra Nevada flow into the lake, and because they escape only by evaporation, the lake water is saturated with salts and minerals. Some of these precipitate into formations called "tufa towers" that, when the water level is low, rise above the surface like open-air stalagmites. Seen against the stark beauty of the Sierra, the shoreline is decidedly unearthly. A good place, it would seem, to seek weird life, and—since the lake water has some of the highest concentrations of arsenic on Earth—especially weird life that likes arsenic.

In August 2009, Wolfe-Simon began working with Ron Oremland, a senior research scientist the US Geological Survey (USGS) and something of an expert in microbes that tolerate arsenic. They collected samples of water and sediment, and Wolfe-Simon carefully cultured bacteria from those samples, gradually and by stages diluting out the amount of phosphorus in their nutrients and increasing the amount of arsenic, with the intention of starving the phosphate users and nourishing the arsenic users, if there were any. By late fall of 2010, she and her research team had concluded that there was at least one arsenic user.

In a paper published in the journal *Science*, and at a NASA-sponsored news conference before a wall-sized image of Mono Lake at its otherworldly best, Wolfe-Simon reported that a bacterium of the family Halomonadaceae used arsenic in many important molecules, including DNA.[16] (She had named it GFAJ-1, an acronym for "Give Felisa a Job"—this an inside joke on anxieties concerning the temporary nature of her position with the USGS and hopes that the discovery might be a career maker.)

The claim was extraordinary, but the evidence for it—at least to many scientists—was less than compelling. They questioned whether the DNA had been sufficiently cleaned, suggested that water would have denatured any (alleged) arsenate-linked DNA, and claimed that remaining traces of phosphorus might have sustained the bacterium's growth. Norman Pace, an internationally respected microbiologist who, with Carl Woese, had done pioneering work on phylogeny and who, as another coauthor of the 2007 NRC report harbored no particular ill will for weird-life research in general, dismissed the work as unworthy of consideration, parsing the blame more or less evenly among "low levels of phosphate in the growth media, naïve investigators and bad reviewers."[17] Shelley Copely gave her own rather devastating take, opining, "This paper should not have been published."[18] There followed a days-long debate among scientists worldwide, much of it carried out in tweets and blog entries, over flaws in the experiment, the problems inherent in scientific peer review, and the general unreliability of NASA's public relations efforts, especially when they concerned microbiology. The paper's authors answered questions in a subsequent issue of *Science* but did not offer to revisit the study, and their critics were unappeased.

The episode became something of an embarrassment for all involved—NASA, whose Astrobiology Institute had supported the work of some of the paper's authors and had sponsored the news conference, the journal *Science* (whose peer reviewers had recommended publication), and of course, the authors themselves. Wolfe-Simon, far from backing away from her claims, has welcomed the critiques as part of the way good science is conducted. As of this writing, the one attempt by other researchers to reproduce Wolfe-Simon's results has failed.[19]

2-Methylamine acid and substitutions (like arsenic for phosphorus) in DNA are only two of the possibilities—informed guesses, as it were—of what to look for if we are seeking weird life in an ecologically separate biosphere. No doubt there are many others, most of them thus far unimagined.

Weird life that is ecologically integrated or biochemically integrated with our own biosphere would prefer less extreme conditions, and so might be more difficult to isolate, but there are ways. Davies suggests we might look for a difference more fundamental than any we've discussed thus far—that presented by the "handedness" of molecules. You can't fit your right hand comfortably into a left-handed glove because the form of the glove is the form of the hand turned inside out, and vice versa. A biologist would say that the two gloves have different *chirality* (from the Greek for "hand"). Large molecules like amino acids and sugars also have chirality, and if they are to fit together to make still-larger molecules, like proteins and DNA, they must have the same chirality.

MIRROR, MIRROR

As it happens, every one of the amino acids used by proteins in familiar life is left-handed, and every sugar in DNA is right-handed. Things didn't have to be this way. Right-handed amino acids would have worked just as well, as long as all were right-handed, and left-handed sugars would have worked just as well, as long as all were left-handed. Things *are* this way because 3.5–3.8 billion years ago, some complex self-organizing prebiotic molecules needing an amino acid happened to use a left-handed one, and some needing a sugar happened to use a right-handed one. The first stitch set the pattern, and it has been followed ever since.

Suppose, however, that in any of the "second genesis" scenarios posited a few pages back, another set of complex self-organizing prebiotic molecules went right where the first had gone left, or went left where the first had gone right. The result might be a sort of weird life that is nearly identical to familiar life but, being made of molecules with mirror chirality, would be unable to interact with it biochemically. How might we find it? In fact, two scientists have already tried.

In 2006, acting on a suggestion from Davies, astrobiologist

Richard Hoover and microbiologist Elena Pikuta put out bait. They began with a standard culture medium, a sort of smorgasbord for microbes, and switched some of its nutrients for their mirror counterparts. Then they took extremophile microbes retrieved from Mono Lake and introduced them to the medium. The researchers expected that, if mirror microbes were living among the extremophile microbes, they would make their presence known by eating the mirror nutrients. Soon enough, something began to eat the nutrients, and after a moment of cautious excitement, Hoover and Pikuta identified that something not as a mirror microbe, but as a heretofore unknown bacterium of the familiar sort, possessed of an unusual ability to chemically alter the mirror nutrients so that it could better digest them. It was a bit of biochemical sleight of hand that Hoover and Pikuta now suspect is owed to certain enzymes. The finding came as a small disappointment, but it was only the first attempt of its kind, and the bacterium that they named *Anaerovirgula multivorans* (roughly, "little rod that will eat anything") was another reminder that a lot of nature was left to discover.[20]

There is another way that weird life might have escaped our attention: by being very, very small.

SIZE MATTERS

Every living organism known is made of cells. Although some cells are quite large (the Gargantua of celldom, *Thiomargarita namibiensis*, is the size of the period at the end of this sentence), most are best measured on the scale of nanometers—a nanometer being one-billionth of a meter. The lower limit for a cell's size seems to be set by *ribosomes*, the (relatively) large molecules of proteins and RNA that work inside all cells to link amino acids and make new proteins. If they are to squeeze ribosomes inside themselves, cells must be a least a few hundred nanometers across. It is for this reason that most microbiologists think that smaller cells are impossible. Most microbiologists—but not all. Benner, for one, has suggested that cells might be much smaller if they made proteins not with ribosomes but with RNA.[21]

There have been at least three reports of very, very small things that—to their discoverers at least—seemed to be living or once living. In 1990, Robert Folk, an emeritus professor at the University of Texas at Austin, discovered in sedimentary rocks tiny structures that he took to be tiny fossils, the calcified remains of organisms a mere 30 nanometers across.[22] He has since found similar structures in other sediments and in meteorites. Some of his peers have been intrigued, and a few have pointed to Folk's findings as evidence that the tiny wormlike formations in the Martian meteorite ALH84001, while much smaller than bacteria, are not too small to have once been living. In 1996, Australian geologist Philippa Uwins was studying sandstone bore samples from a deep-ocean borehole off the coast of western Australia. She and her colleagues found tiny filaments that under an electron microscope looked like blobs in a lava lamp. By an ingenious method, Uwins was able to show that there was DNA inside the structures (not just on their surfaces)—evidence that they might be living, or at least might once have been living.[23] In 1988, Finnish biochemist Olavi Kajander was examining cells with an electron microscope and found within them tiny particles some 20 nanometers across.[24] Believing them to be living, he called them "nanobacteria." Of the three discoveries, Kajander's may be the strangest—and the most unsettling—because the particles are found in human tissue.

At present, the preponderance of evidence is that none of

these findings is an organism, living or once living. In 2003 a research group concluded that what Folk had found were probably nothing more than by-products of bacteria with rather more typical dimensions.[25] Recent research suggests that Uwins's filaments are calcium carbonate and organic material that at some stage of development had encapsulated pieces of DNA.[26] And a National Institutes of Health (NIH) study published in 2000 threw serious doubt on Kajander's claims, which had already come under fire from several quarters. It should be said that Kajander himself continues to believe his "nanobacteria" are living, and his second thoughts are limited to his choice of nomenclature; he recently said that he probably should have given his findings a less provocative name, like (this is his phrase) "calcifying self-propagating nanoparticles."[27]

Most microbiologists have given these findings a wide berth. One reason is that the work done so far—Kajander's in particular—has generated controversies that give pause to scientists concerned for their careers and reputations. Nanoparticles, it seems, fall into a disciplinary no-man's-land between chemistry and biology. Microbiologist John Cisar, who led the NIH study that countered Kajander, noted, "I'm not saying there's nothing there. It's just that we were looking at it from a microbiologist's perspective. And when we didn't find any signs of life, we moved on."[28] These findings may represent a class of forms somewhere between nonlife and life, forms unknown to science. But whatever they are—weird life, unusual chemistry, or something in between—very small weird life remains a real possibility. As David McKay noted of Uwins's discovery (and it might easily apply to the others), "It's something that shows that we just do not understand the small end of the spectrum."[29]

As should by now be obvious, the big challenge facing seekers of weird life is that it could be weird in any number of ways, most of which we haven't thought of. It is for this reason that Davies has suggested that the strategy with the best chance of success is simply to broaden our gaze and look for things that are unexplained. One such thing, of particular interest to Carol Cleland, has been unexplained for a very long time.

DESERT VARNISH

In 1832, a young Charles Darwin was acting as assistant to the captain and unofficial naturalist aboard the HMS *Beagle*. As the ship was anchored off the coast of South America near San Salvador, Darwin explored the shore, where he was intrigued by rock outcroppings that glittered in the sunlight and seemed "burnished." He deduced that the rocks shone because of a coating of thin layers of metallic oxides, but he could not explain how it might have been made.[30]

Geologists have since found the same coating—now called "desert varnish"—in many locales. Although they are no more certain of its provenance than was Darwin, they have ideas. They also have two observations that give reason to suspect that this substance may be a product of biology. The first is that the thin layers of minerals and chemicals that desert varnish reveals in cross section resemble the layers found in "stromatolites." These are the mineral formations that in Shark Bay, Australia, look like half-submerged tortoise shells and in upstate New York look like fossilized cauliflower. Stromatolite layers are formed by generations of bacteria that, like a medieval city built and rebuilt on its own ruins, lived and died one atop another. The layering of desert varnish, so the thinking goes, might result from a similar biological process. The second observation is that many of the chemicals in desert varnish layers, most notably manganese and iron, are not in the rocks (like sandstone) that desert varnish typically varnishes, but are in fact produced by known organisms.

Even together, though, these observations are a long way from a clear-cut case for life. No laboratory microbiologist has been able to coax bacteria or algae to make desert varnish. And just as discouraging (to those who might wish it to be a product of organisms, that is), bacteria found in *in situ* desert varnish (the bacteria that might reasonably be expected to produce it) are of many varieties—too many, microbiologists think, to turn out the same product so consistently.

It is possible—and this is the prevailing view—that the stuff that intrigued Darwin and so many after him is the end result of some very complex chemistry. But no one has been able to reproduce that either. And so we have it: a natural phenomenon that exists in plain sight, and that after nearly two centuries of study remains a mystery. It is, so Cleland thinks, a fair candidate for weird life.

So far, we've learned of extremophiles that live at the outer boundaries of life as we know it. We've also learned of possibilities for life on Earth that, by subtle and not-so-subtle differences in their metabolism, might live beyond those boundaries. But we've been hugging the shoreline and reconnoitering a few nearby islands. Much of the remainder of this book will describe ideas of life that is much, much weirder. We'll venture into waters that are little charted, and are sometimes out of sight of the shoreline altogether. Before we do, though, we would be prudent to establish exactly where that shoreline is, and to take a good look back at it.

CHAPTER THREE

Defining Life

More than once in recent years, planetary scientists have been surprised to find water where they didn't expect to find it. On Mars, for example. The planet's atmosphere is so thin—so near a vacuum, in fact—that any liquid water on its surface would evaporate immediately, and even ice would be likely to sublime directly into water vapor. But since the mid-1970s, when NASA's twin *Viking* spacecraft imaged dry riverbeds and meandering tributary channels, it became apparent that once upon a time, perhaps 4 billion years ago, the surface of Mars had seen cataclysmic flooding. In the first decade of the twenty-first century, a small armada of spacecraft found more evidence of a watery Martian past—an outcropping that may be the shore of an ancient sea, and pack ice covered in volcanic ash. Most surprising were dry gullies and streambeds formed only a few thousand years ago, and evidence of a flash flood that happened even as the reconnaissance was under way. Since the 1970s, scientists had believed that when the planet lost most of its atmosphere, much of its water went with it. Now they know that some held on as permafrost, and still more remains in liquid form beneath the surface.

Astrobiologists have long noted that water on the surface of a planet too close to a star will boil; water on the surface of a planet too far away will freeze. Naturally, there is a slender range of distances from a star at which water could exist on the surface of a planet or moon—just the right distances at which water is liquid, the state necessary to life as we know it. It's called the *habitable zone* or *Goldilocks zone*—the second because, like the porridge favored by the girl in the fairy tale, it is neither too hot nor too cold, but "just right." It turns out to be a shell-shaped section of space whose inner surface is just inside Earth's orbit and whose outer surface is just beyond Mars's.* A lot of space, we might say. But when we see it pictured within the much larger volume of the whole Solar System, we realize how small a region it is. Although there is a lot of porridge in the universe, very little of it is just right.[1]

Or so it once seemed.

———

* Traditionally, the habitable zone for any planet (or moon) is the region where water would be liquid on its surface. For a planet whose atmosphere is chemically active, the zone's inner boundary is the distance from its sun at which it is so warm that water vapor accumulates in the upper atmosphere, where unfiltered ultraviolet radiation splits the molecules into its components: hydrogen and oxygen. The hydrogen escapes into space, and the oxygen is eventually absorbed into surface rocks. The planet dries up, a runaway greenhouse effect begins, and soon it is too warm for liquid water. The outer boundary for a planet or moon whose atmosphere is chemically active is the distance from its sun at which carbon dioxide, a greenhouse gas, freezes out of the atmosphere and the planet becomes too *cold* for liquid water. As it happens, Mars orbits within our Solar System's habitable zone; liquid water once existed on its surface not because the planet was nearer the Sun (it wasn't), but because that surface benefited from the greenhouse effect of a more substantial atmosphere, mostly of carbon dioxide. (Hart, "Habitable Zones")

WATER, WATER EVERYWHERE

In 1995, NASA's *Galileo* spacecraft detected magnetic fields near three of Jupiter's large Galilean moons—evidence that miles beneath their frozen surfaces are briny oceans, warmed by natural radioactive heating from their cores and, in the case of Europa, the tidal heating caused by the push and pull of gravity. Some suspect that Europa's ocean, beneath a crust of ice tens or hundreds of meters thick, may hold twice as much water as that in all of Earth's oceans combined.[2] Planetary scientists have reason to believe that great reservoirs of water-ammonia mixes are also held beneath the surfaces of Callisto and Ganymede.

Scientists have evidence for water inside planets and moons in still-colder regions, kept in liquid state by pressure, tidal heating, radioactive decay, or some combination of these. Adam Showman, a planetary scientist at the University of Arizona who has modeled interiors of moons of the Solar System's outer planets, estimates that at least twelve hold some liquid water.[3] In a 2006 paper, German physicist Hauke Hussmann and his colleagues modeled interior structures for medium-sized icy worlds in the outer Solar System, assuming heating in their cores from natural radioactivity, and concluded that subsurface oceans are feasible on Rhea, a smaller moon of Saturn; Titania and Oberon, moons of Uranus; Neptune's moon Triton; and the dwarf planet Pluto.[4] As to worlds beyond the Solar System—in the last several years NASA's Kepler space observatory has made provisional identifications of more than 2,000 candidates for planets, almost fifty of which are in their stars' habitable zones.

It is not difficult to imagine Earth's extremophiles doing well in at least some of these places—the intraterrestrials beneath the surface of Mars, hydrothermal vent communities in Europa's vast dark ocean, and Antarctic single-celled algae in the icy fissures of Saturn's moon Enceladus. In fact, some have suggested that if and when humans decide to "terraform" Mars—that is, initiate a centuries-long process through which Mars might be provided a breathable atmosphere and balmy temperatures—they might begin by seeding it with a few especially hardy extremophiles.[5]

All this water has been very good news for NASA, because the agency's strategy in its search for extraterrestrial life, informally termed "follow the water," has long assumed that any habitable extraterrestrial environments must have liquid water. It's a reasonable assumption, especially given current knowledge of biochemistry and the limited resources of a government agency whose funding depends, after all, on taxpayers. And as we now know, there are certainly plenty of places to look.[*] Nonetheless, the authors of the NRC report suspected that following the water might limit any discovery to organisms like those we know. And they began their work contending that if NASA scientists expected to look for life in *all* the places it might be possible, and if they expected to recognize it when they (or their instruments) saw it, they needed a proper definition of life—that is, a definition open to all possibilities, yet rigorous nonetheless.

If you thought that biologists might have an all-purpose defi-

---

[*] "Follow the water" is the most recent iteration of a long tradition with a somewhat dubious beginning. Astronomer Percival Lowell observed the planet Mars for many nights over several years, and imagined that he saw crosshatched dark lines that he believed to be vegetation growing alongside a planetwide network of canals. From such figments he produced astonishingly detailed maps and hypothesized a civilization that inspired the "Martian" science fiction of H. G. Wells, Edgar Rice Burroughs, and Ray Bradbury. (Bradbury et al., *Mars and the Mind of Man*)

nition at the ready, you'd be wrong. There are at least nine specialties within biology, and biologists tend to define life according to their own.* A physiologist might call life "a system capable of eating and metabolizing"; a molecular biologist might call it "a system that contains reproducible hereditary information coded in nucleic acid molecules." Seeking a broader definition, we might look to a field of study with a wider view—say, philosophy. In fact, philosophers have been formulating definitions of life through much of recorded history. Most have characterized life not according to what it is made of (until the late nineteenth century no one had any idea), but according to what it does. And most have come up short. The problem is that any reasonably complete list of an organism's functions is bound to include some that are performed by things that are nonliving, and—just as problematic—some nonliving things perform functions that some living organisms cannot perform.

For instance, suppose we define a living organism as that which grows, consumes, converts matter into heat energy, maintains a metabolism (that is, perpetuates itself through chemical activity), and after a fashion dies. Candle flames and stars do all these things, and few would call them living. So we reconsider and define a living organism as that which, in addition to the previous, reproduces itself. Problem solved, we think. Until we recall that crystals reproduce, and most of us do not count them as living. And we remember that mules and worker ants cannot reproduce, yet they are to all appearances alive.

—————

* Biological specialties include at least these nine: anatomy, taxonomy, physiology, biochemistry, molecular biology, ecology, ethology, embryology, and evolutionary biology.

A PARTS LIST FOR LIFE

Attempts to define life by what it is or what it is made of have proved at least as difficult. The answers, in the broadest historical terms, have appeared as part of two suppositions. One was that any organism is animated by something that, like a spirit or soul, cannot be weighed, measured, or seen—something that came to be called a "vital force." The idea may strike us as redolent of druids and tree spirits, but scientific arguments for such a force lasted into the early twentieth century. The other supposition—that life is fundamentally material—was well articulated in 1868 when Darwin's staunch proponent T. H. Huxley declared that all organisms are composed of chemical compounds that are themselves lifeless.[6]

In the late nineteenth century and early twentieth, the materialist view gained much support. But since no one could be sure exactly *which* chemical compounds compose organisms, the case was never quite closed, and a strict definition of life remained elusive. In 1937 the British biologist Norman W. Pirie claimed that the terms "life" and "living," especially as applied to cases for which the transition from nonliving is so gradual as to have no discernible boundary, were worthless.[7] In a 1944 treatise called *What Is Life?*, physicist Erwin Schrödinger was more hopeful. He maintained that eventually life would be defined in some detail by physics and chemistry. But as years and decades passed, eventually was beginning to seem like a long time. As late as the 1960s, a much-respected textbook all but advised surrender, remarking that "attempts at an exact definition of life are not only fruitless, but meaningless."[8]

A DARWINIAN DEFINITION

Well before the 1960s, it had become possible to add another item to a list of life's functions, one that—conveniently for list makers—collapsed most of the others into itself. Life evolves, and any living organism is by definition a product of evolution. (In fact, by 1973 Darwin's basic insight had been so thoroughly confirmed that Ukrainian-American geneticist Theodosius Dobzhansky could title an essay "Nothing in Biology Makes Sense Except in the Light of Evolution.") Norman Horowitz, the geneticist and professor of biology at Caltech who in the 1950s had discovered that one gene governs one enzyme, elaborated. "Living organisms," he wrote, "are systems that reproduce, mutate, and reproduce their mutations."[9] When those mutations are subject to natural selection over time, the organism adapts to its environment with ever-increasing precision.

If we define life as that which evolves, however, we are met immediately with another problem: viruses. *Viruses* are like nothing else in our category of the living. They are one-tenth to one-hundredth the size of the smallest bacterial cell, and nowhere near as complex as that cell, being little more than sets of genes held within a capsule of protein molecules. Viruses neither eat nor excrete, and as parasites they can do their reproducing only within the cell of an organism. It is this last behavioral limitation that compels authors of most biology textbooks to treat viruses as nonliving, and allows high school students reading those textbooks to discover, on a perennial basis, a scientifically grounded addition to their lexicon of insults.[10] Yet it so happens that viruses evolve—which is to say they reproduce, mutate, and experience natural selection.

Suppose Horowitz is right that evolution is life's defining feature, and that viruses are living. Or suppose that although evolution is an essential feature of life, it is not the defining feature, and that viruses (lacking that defining feature or set of features) are nonliving. In either case we still face a problem. There may be things that are nonliving yet evolve in the literal sense of that word. I'm writing this passage in a university library, a place housing journals, computer terminals, electronic databases, and of course, millions of books. A case might be made that all are as much products of evolution as the librarians who so kindly answer my reference questions. Consider the books. Some, like wildly successful species, enjoyed large press runs and many editions, producing, one might say, several generations of ever-larger populations. Others existed for a single edition of a few dozen copies, one of which happens to be preserved here like a rare fossil in a natural history museum. In marketplace competition, there were winners and losers. Some pushed into new markets entirely, and in so doing made a niche for their successors. A few, like a brave or unwitting microbe beginning a phylum, established what would come to be recognized as a new literary genre.

Biologist Richard Dawkins would demur a bit, maintaining that books are not themselves products of evolution. Rather, he would say, they are the vehicles that enable the evolution of ideas, stories, and language—these being examples of the units of cultural transmission and replicating entities that he calls "memes." Arguing that evolution is "too big a theory to be confined to the narrow context of the gene," Dawkins claims that memes evolve, and do so in the full sense of that word.[11] Further, he maintains that since the ability to evolve is a property exclusive to living things, it follows that memes are living—not merely metaphorically, but literally.[12] Many have made the same case for certain computer software programs, the first and most famous of which is British mathematician John Conway's "Game of Life," which used a few simple rules to generate complex self-organizing patterns.[13]

For all this, some claim that evolution may *not* be life's defining feature, their evidence being certain organisms that are manifestly living but, in their view, did not evolve. They have argued, for instance, that the natural selection of many plants was suspended with the invention of agriculture, and the natural selection of humans was suspended with the invention of medicine.[14] An evolutionary biologist might counter that no organism is beyond the reach of natural selection—that a highly cultivated plant like the tulip has insinuated itself into a relationship with another species (our own) that ensures the survival of its line, and that persons whose life spans are increased by medicine might possess genes that elicit sympathy or altruism in others (eventually producing medical students, physicians, and a good bedside manner), which in turn benefit other members of the species.

Even if award-winning tulips and children of parents with excellent health care are products of evolution, it is certainly possible to imagine organisms made of chemical compounds that do not evolve and yet—because they might metabolize and reproduce themselves—would justifiably be called living.[15] It is also possible, incidentally, to imagine evolution by different means—for instance, evolution whose mutations are provided not by RNA and DNA, but by random errors in the synthesis of chemicals used to replicate and metabolize.[16] We would likely call them "alive" as well.

If anything is clear from all this, it is that any category of things called "living" will have exceptions, and any reasonable definition of life is likely to be provisional. So it isn't surprising that the NRC report hedges on the matter. It lists the characteristics of known life and allows that the better definitions are those along the lines of a "chemical system capable of Darwinian evolution."[17] But it does not offer a new definition and, in so many words, admits that any such attempt seems doomed to fail. Where does the matter stand now? In some ways, about where it stood three-quarters of a century ago, when Schrödinger posed the question.[18] It seems that natural philosophers and scientists have again and again made provisional definitions that seemed at first satisfactory but, on further examination, were found to have boundaries that were shifting and indistinct. Meanwhile, the thing they were trying to define—amoeba-like—slid, slipped, and wriggled free.

A census taker, standing in a dimly lit hallway, knocks on a door. It is opened by a man in a bathrobe, a bit disheveled. The census taker introduces himself, and the man agrees to answer questions. The census taker begins: "Do you receive mail at this address?" "Do you pay utility bills for this apartment?" "Do you pay rent to the owner of this building?" The man answers "Yes" to all of these. Finally, the census taker asks, "Including yourself, how many people are living here?" The man says, "None." At this the census taker says, "I don't understand. You receive mail here, pay utilities here, and pay rent here—but you aren't living here?" The man opens the door wider, allowing a view of worn furniture and a threadbare carpet, and shrugs. "*This* you call living?"

It's usually poor form to explain a joke, but I'll risk impropriety in the service of a point and observe that some of its humor derives from the census taker's assumption that a definition of living could be met with a list of criteria. Perhaps we've been acting a little like him. The problem is not merely that life resists our efforts to define it; it is that even a good definition (if we had one) would be nothing but a list of features, lacking an underlying principle explaining how those features are related or why they should appear together. A definition, in other words, is of little help in understanding life.

If we are to understand life, we need a *theory*—that is, a set of self-consistent hypotheses for defining *all* life, familiar and weird—that makes predictions and can be tested. Recently, Carol Cleland made the same point, noting that we have been defining life by enumerating features, imagined or real, and—like medieval philosophers—bickering over which are significant. She notes that without a molecular theory we would still be defining water as that which is wet, or as a colorless, odorless liquid.[19] We were able to agree on a universally applicable definition of water (two hydrogen atoms covalently bonded to a single oxygen atom) only when we had a theory of matter at the molecular scale.

A THEORY OF LIFE

To construct a theory of life we will need to know which features in living organisms are necessary and which are merely contingent. Of course we can make guesses, but to be certain as to which is which we will need a second example of life. The realization is not new. Half a century ago, a panel of leading biologists commissioned by the US National Academy of Sciences opined, "The existence and accessibility of Martian life would mark the beginning of a true general biology, of which the terrestrial is a special case."[20] The panel argued that the chance to develop a theory of life was in itself a reason to explore Mars. Of course, fifty years later we still do not know whether Mars holds or once held life. We have no second example and no theory.

Can we lower our sights further, and at least draw up a set

of criteria that scientists can use to point to something and say, "That's alive"? In fact, NASA scientists asked themselves the same question—again, a half a century ago.

In 1960, America's newly formed space agency was as ambitious as it was busy—especially with regard to questions relating to life elsewhere. In that single year it created an Office of Life Sciences to consider biological issues related to the exploration of neighboring worlds (including, interestingly enough, the danger of contaminating those worlds by Earth microbes); it established a life sciences laboratory at its Ames Research Center to study the conditions under which life might survive; and it authorized the Jet Propulsion Laboratory (JPL) in Pasadena, California, to consider the type of spacecraft that might be employed to search for life on Mars.

The early NASA was nothing if not forward-thinking. With unmanned flybys of Mars still years away, the agency had already contracted designs for instruments intended to detect life on its surface. Those instruments would be carried by two unmanned spacecraft. Their first iteration, a program for an especially ambitious remote-controlled laboratory called *Voyager*, was canceled in 1967.* The second, a scaled-back version approved by Congress a year later, was called *Viking*. For both programs, the clearinghouse for life-detecting proposals was the office of project scientist Gerald A. Soffen at JPL.

Among those attending Soffen's brainstorming sessions was a forty-year-old British inventor of medical equipment named James Lovelock. Lovelock couldn't help but observe that most of the proposals were of the "add water and stir over low heat" vari-

---

* The same name would be given the twin spacecraft that in the 1980s conducted enormously successful flybys of the Solar System's outer planets.

ety, betraying rather parochial assumptions about the nature of life. Some in the session challenged Lovelock to describe a way to cast a wider experimental net, and he said that he would look for "thermodynamic disequilibrium," a condition in which otherwise inert matter channels energy into a form that counters entropy—that universal tendency of things to slide, slip, fall down, and fall apart. At that moment Lovelock had given no thought to exactly how one might detect thermodynamic disequilibrium on Mars, but in the days following he reread Schrödinger's treatise and soon came up with a short list of approaches. The most promising, he thought, would be a chemical analysis of the planet's atmosphere.[21]

For an atmosphere in thermodynamic equilibrium, the chemical transactions are over and the scores are settled. All the chemistry that can happen *has* happened. But an atmosphere in disequilibrium is unsettled, and chemical transactions continue. Lovelock's point was that if you detected the presence of a highly reactive chemical, one absorbed quickly and efficiently by other chemicals in the environment, you would have reason to suspect that it was being produced just as quickly and efficiently— perhaps, as is the case with oxygen and methane on Earth, by living organisms.

While at JPL, Lovelock met a young planetary scientist named Carl Sagan. Sagan disagreed with Lovelock on several counts, but he was intrigued enough to publish Lovelock's work in *Icarus*, a journal he edited, and somewhat later to introduce him to Lynn Margulis (Sagan's former wife), a biologist who was interested, like Lovelock, in the unexplained stability of the chemistry of Earth's atmosphere over geological time. In subsequent years and with Margulis's help, Lovelock would develop the notion that thermodynamic disequilibrium signifies life into the Gaia hypothesis and theory. Both are propositions that the Earth and its life compose a single complex system analogous to a self-regulating organism, and that through various feedback loops the system regulates (and for several billion years *has* regulated) the chemical composition and temperature of its atmosphere and oceans so as to keep them suitable for life.

Soffen's office considered fifty proposals for various life-detecting instruments. Because of size and weight constraints, only three were chosen. Lovelock's was not among them.[22]

Each of the two *Viking* landers had a mechanical sampler arm with a scoop on its end. It would use this arm to dig soil and drop it into one of three hoppers. From these the soil would be dropped below decks into the "biology package," this being a collection of pipes, hoses, and small tanks of gases and incubation chambers that rotated on carousels. The package, weighing all of twenty pounds and squeezed into a cubic foot, represented what one historian called "the most sophisticated thinking of the twentieth century on the subject of extraterrestrial life in the solar system."[23] Nonetheless, the three experiments it contained were designed around a simple and straightforward premise: that any organism would take in nutrients and discharge waste.

The "labeled release" experiment was developed from methods of detecting contaminants in municipal water supplies by Gilbert Levin of Biospherics Inc. In its Martian version, a dilute solution of radioactively labeled organic compounds would be added to an incubation chamber containing a few ounces of soil. If organisms ate the compounds and exhaled gases like carbon dioxide, hydrogen, or methane, those gases would be identified by an onboard carbon-14 detector.

The "gas exchange" experiment, designed by Vance Oyama of

NASA's Ames Research Center, would add water moisture (or, in another mode, a rich broth of organic compounds) to a soil sample, and also would test for waste gases—not with a carbon-14 detector as in Levin's experiment, but with a chromatograph.

The "pyrolytic release" experiment was developed by Caltech's Norman Horowitz, mentioned earlier, in the discussion of the definition of life. It would use radioactive labeling as did Levin's, but (at least in Horowitz's view) it would make fewer suppositions about Martian life. The only things Horowitz's experiment would add to the soil in the incubation chamber were carbon dioxide and carbon monoxide, gases known to exist in Mars's atmosphere.[24]

Levin and Oyama suspected Horowitz was offering so little that he might fail to provoke a metabolism, but Horowitz was sure that one thing the Martians would *not* want was water, since on Mars's surface, the environment in which they had presumably evolved, it would freeze or evaporate. Horowitz thought that his experiment was the only one that was purely Martian, and that Levin and Oyama were looking for Earthlike life on Mars. Ideally, the incubation chamber that his experiment used would mimic Martian conditions, and it particularly irked him that in order for Levin and Oyama to be able to use liquid water, the temperature of the whole biology package—including his incubation chamber—was kept at 10°C, some 60 degrees higher than the summer average for the places the *Viking* landers would touch down.[25] In the end, the fact that the twin landers could not carry scientists may have been for the best. It would have been a very long trip.

SQUAMOUS PURPLE OVOIDS

The three *Viking* biology experiments were part of thirteen separate investigations conducted by as many teams of scientists—seventy-eight researchers in all. Among them was Sagan. In 1975, the year of the *Viking* launches, he was forty-one, the David Duncan Professor of Astronomy and Space Sciences at Cornell, and author of several popular books about astrobiology. Sagan was a frequent guest on TV news programs and talk shows, where he proved a refreshing antidote to stereotypes of the scientist as drone. He was witty, charming, and above all else, enthusiastic—visibly excited about what he was doing.

Because much of Sagan's scholarly work, like his popular writing, speculated on the nature of extraterrestrial life (indeed, his ideas of recurrent warm, wet periods on Mars and microbes taking water from rocks now seem prescient),[26] anyone might have expected him to be included on *Viking*'s biology team. He was not. Instead he was a member of the "imaging" team, the group that would operate the cameras (two on each lander) and analyze the images returned. *Viking*'s designers gave it cameras so that the geology teams might study the surrounding terrain. There was some irony in Sagan's position, as he had once attracted no small amount of ire from a conference of lunar geologists when he called the Moon "dull," and he was not particularly interested in geology except in cases where it might be associated with life. This happened to be one. Sagan opined that the cameras might do double duty as *Viking*'s fourth biology experiment, one that made fewer suppositions about the nature of its subject than even Horowitz's. It did make one supposition, though, and it was far from insignificant: that Mars might have organisms large enough to be seen without a microscope.

Most of the biologists associated with choosing and planning the life-detecting experiments aboard *Viking* (and *Voyager* before it) assumed that any Martians would be microbes. There were two justifications for this thinking. First, microbes make up most of Earth's life, and until quite recently (on geological timescales) they made up all of it. If our planet's history is any guide, then microbial life is the first sort to arise, and large, complex organisms appear much later if indeed they appear at all. Second, scientists' knowledge of Mars's surface, gained from Earthbound observations and the *Mariner* series of spacecraft, showed an environment far more foreboding than the interior regions of Antarctica, where what little life that survived was microscopic.

Nonetheless, Sagan confessed a real worry that while scientists had their remote-controlled noses stuck in a few ounces of Martian soil, something big (in his own coinage, a "macrobe") would crawl, scamper, or flitter by unnoticed. It concerned him that *Viking*'s cameras, called single-slit scanners, worked so slowly—with a mirror moving up and down behind a narrow vertical slit in a housing that pivoted by increments—that they might miss something that was moving quickly. So, during a test in the Mars-like landscape of Colorado's Great Sand Dunes National Monument, Sagan borrowed a garter snake, two turtles, and a lizard from a pet store and placed them before the scanner. The camera caught the snake and turtles as a blur, and the tracks of the lizard. Sagan was reassured. Even if the cameras missed the macrobes, they could detect their evidence.

In interpreting that evidence, however, Sagan acknowledged a difficulty. "Suppose," he said, "a fourteen-foot squamous purple ovoid with thirty tentacles came floating through the air, and Viking got a picture of it. We'd identify it as alive even though we'd never seen it before and didn't know its chemistry simply because it was so improbable."[27] In other words, a squamous purple ovoid would obviously be more than a happenstance collection of inert matter. It would be—and here Sagan borrowed a page from Lovelock—in a state of thermodynamic disequilibrium, and anything in thermodynamic disequilibrium stood a good chance of being alive.

Sagan suggested that the thermodynamic disequilibrium of organisms that were macroscopic and surface-dwelling might be obvious. They would look top-heavy. Like dandelions, cows, and dairy farmers (or examples of each when standing, anyway), they would present a greater surface area farther from the ground than they would nearer the ground. Top-heaviness was hardly a guaranteed biosignature; and as if to make that point, the leader of the imaging team, geologist Timothy Mutch, displayed in his office photographs of ventifacts, enormous rocks shaped like fluted vases, their concave surfaces abraded by windblown sand. But it was, Sagan maintained, a good place to start.

THE VIEW FROM MARS

In July 1976, *Viking 1* made a successful landing on the low-lying Chryse plain in Mars's northern hemisphere. Almost immediately its single-line scanners whirred and clicked, and images of the surrounding landscape were radioed Earthward. They showed salmon-colored sand dunes stretching to the horizon and strewn with sharp-angled rocks. It was the first time anyone had seen a landscape from another planet, and that landscape was hauntingly beautiful. It was also, to all appearances, lifeless. "There was," Sagan said, "not a single recognizable funny-looking thing, no obvious sign of thermodynamic disequilibrium."[28] The biologists were intrigued briefly by certain rocks, and for a few days Sagan wondered about a roughly spherical one, but it was clear that except for some windblown sand, nothing within view of the lander was moving.[29] The images sent from *Viking 2*, which landed some weeks later, showed a landscape just as barren, just as still. Or, as they say in the movies, was it?

Although Oyama and his team concluded that the oxygen produced in his "gas exchange" experiment had resulted from a chemical reaction involving hydrogen peroxide in the soil, the other two biology experiments (Levin's and Horowitz's) gave "presumptive positive results." But *only* presumptive, and difficult to have confidence in because they were contradicted by a separate experiment: a molecular analysis of the soil, in what was later called "probably the most surprising single discovery of the mission,"[30] failed to detect organic compounds.

Immediately, each member of the biology team dropped his carefully planned experimental strategy and put all energies into determining whether the reactions were biological or (merely) chemical. This would have been a fair challenge with the samples sitting in a fully equipped Earthbound laboratory, but it was particularly daunting with samples more than a million miles distant in what was, after all, a rather modest facility. Still, they soldiered on, and nine months later, with the experiments put through twenty-six separate cycles, the results were in. That was the good news. The bad news was that no one agreed on what they meant.

Oyama affirmed his earlier, nonbiological take, and Horowitz now drew a similar conclusion. Horowitz had induced changes in temperature experimentally, and he reasoned that if the reaction he measured had been biological, it would have been more sensitive to those changes. Levin had held from the start that his "labeled release" experiment detected life, and after subsequent cycles he felt only more certain.

Some said the confusion would have been avoided if NASA's first attempt to detect life on Mars had not resorted to a million-mile Hail Mary pass, but instead had proceeded in a more methodical, stepwise fashion so that scientists could make proper studies of the chemistry of the soil and atmosphere before they designed—let alone conducted—tests for living organisms. Others thought the biological tests themselves might have been better coordinated, made more like the preliminary plans for *Voyager*, which had proposed thirty separate life-detecting experiments to be performed in sequence, each addressing questions raised by the one that came before. Still others countered that the *Viking* experiments *were* teaching us about Martian chemistry, and giving results so unexpected that no one would have deliberately designed experiments to test for them. In the end, if the confusion proved anything, it was that, as Sagan noted, "looking for life is hard."[31]

In 1978, Levin published an article claiming that images from Chryse, on further study, revealed greenish patches on some rocks, and that the patches moved. Neither his colleagues nor the imaging team saw green; most suspected that whatever Levin saw could be attributed to shadows or dust. In the years immediately following, the consensus of scientific opinion was that, Levin's objections notwithstanding, *Viking* had not found life on Mars. Whether there *was* life on Mars, strictly speaking, remained an unanswered question, but most who studied the results were doubtful. Horowitz made a small show of appeasing the few holdouts, noting that it was impossible to prove that the reactions measured were *not* biological, but only in the way it was impossible to prove that the rocks surrounding each lander were not organisms that happened to look like rocks.[32] Soffen was more blunt. "I began with an optimistic view of the chances of life on Mars," he said. "I now believe that it is very unlikely."[33]

Thirty years later, however, this unhappy consensus was upended by two discoveries occurring a million miles apart: evidence of liquid water on Mars and intraterrestrial microbes like *Bacillus infernus* on (or rather, in) Earth. At present, we still don't know whether life exists on Mars—but we now know that physics and chemistry allow its possibility.

Against these shifting views, Gilbert Levin's have changed only slightly. Now an eighty-something adjunct professor working alongside Paul Davies, Levin is redesigning his "labeled release" experiment to identify microbes in extreme environments on Earth. As to his putative Martian microbes, their time may have come at last. Astrobiologist Dirk Schulze-Makuch of Washington State University and Joop Houtkooper of Justus Liebig University (in Giessen, Germany) recently took a second look at the *Viking* results and, following a line of thinking begun by Horowitz, they hold that Martian microbes—or those living on the surface, at least—would have no experience or need for liquid water, where it would freeze or sublime. But they might know and love a liquid with a lower freezing point, like hydrogen peroxide. The chemical is fatal to many terrestrial bacteria (this is the reason it may be in your medicine cabinet), but when mixed with the right compounds it can be tolerated and can actually assist functions inside cells.

Schulze-Makuch and Houtkooper claim that *Viking*'s molecular analysis of Martian soil may have failed to detect organic compounds (that bewildering result) because the hydrogen peroxide released from dying cells oxidized those selfsame compounds. In fact, they contend, microbes that use hydrogen peroxide could have produced nearly all the results of the *Viking* biology experiments. In Levin's experiment, gas was produced when the nutrient was added, but the reaction tapered off—generally speaking, not a biological result. Schulze-Makuch and Houtkooper suggest that the tapering off represented the last gasps of microbes that, exposed to the experiment's liquid water, had after a few seconds drowned or burst.

The researchers' hypothesis, even untested, is intriguing and perhaps unsettling. Intriguing in that we may have already found extraterrestrial organisms, and weird ones at that. Unsettling in that inadvertently, and with the best of intentions, we may have killed them.

CHAPTER FOUR

Starting from Scratch

F ifty years after the *Viking* planning sessions, scientists are again asking themselves whether they could be certain to recognize life if they or their instruments saw it. Were they obliged to use the NRC report's provisional definition of life as a "chemical system capable of Darwinian evolution," their answer would be no. Chemical systems are common, and although evolutionary biologists can point to innumerable examples of evolution over time, they find evolution in action difficult to identify even on Earth. Identifying a system *capable* of evolution might be easier, but only slightly.

So, scientists have no definition or theory of life and, since biosignatures are evidence but not proof, no set of criteria by which they can be sure to recognize it. Can they say anything about life that might assist in a search for its weirder versions? Perhaps. A list of what life needs in the way of materials and conditions might allow a search strategy wider than "follow the water," yet still selective. It would tell scientists where they might look, and just as important—considering limits on time and resources—where they needn't bother to look.

WHAT LIFE NEEDS

If life has a fundamental, nonnegotiable, rock-bottom requirement, it is a source of energy. Familiar life takes energy from sunlight, chemical reactions, thermal energy, and natural radiation. It uses these sources in particular, biologists presume, because they are abundant and freely available. (Interestingly, there are other sources—ultraviolet radiation, thermal gradients, and gravity—that, as far as anyone knows, familiar life does *not* exploit, and exactly why is far from clear.)[1] In addition to an energy source to drive its chemistry, life needs a medium in which that chemistry can work—that is, a medium in which molecules are suspended and through which they can move freely and interact easily. If we are in a mood to be open-minded, this requirement may be negotiable; many science fiction writers and some scientists have imagined organisms made of atomic nuclei and magnetic fields that might survive in a vacuum, and I'll discuss these in a later chapter. For the moment, though, let's stay conservative, follow the NRC report's provisional definition, and assume that all life is based in chemistry and needs a medium.

There are a great many mediums, and some are rather exotic. I'll be prudent here too, and begin with the mediums we know best—those represented by the three "classical" states of matter. They are, of course, solids, liquids, and gases. As potential staging grounds for life, solids look like a bad bet because they don't allow molecules enough movement. Gases look like a bad bet because they don't allow molecules enough contact. But liquids, the happy medium of mediums, allow plenty of both. Recalling life's great affinity for water, we might suppose that for familiar life, water is the liquid medium of choice. But the fact is—this a point made by several astrobiologists—life probably *had* no choice. As Dirk Schulze-Makuch observes, "Life on Earth learned to work with water because it's the only liquid that's really abundant."[2] In fact, in extolling the virtues of water a few chapters back, I may have been giving credit where it wasn't quite due. From its first appearance in the remnants of cooling supernova explosions somewhere in a young universe many billions of years ago, to the steady drip of your leaky kitchen faucet, water hasn't changed. A molecule of water was an atom of oxygen covalently bonded to two atoms of hydrogen then, and a molecule of water is an atom of oxygen covalently bonded to two atoms of hydrogen now. But life, since its first appearance, has changed enormously, and no doubt some of those changes took advantage of water's unusual characteristics—its surface tension, the wide range of temperature at which it is liquid, and so forth. In other words, life learned to love the one it's with.[*]

The truth is that astrobiologists can conceive of a variety of liquids that might serve as mediums for life, each with characteristics that life might use to its advantage. Water carries around hydrogen ions that catalyze chemical reactions crucial for cells to metabolize nutrients, but a number of other liquids—among them hydrogen fluoride, sulfuric acid, ammonia, and hydrogen

---

[*] Water, however, has not responded in kind. Since water is a nonliving and nonevolving chemical compound, we should not be surprised. We might be surprised, though, to learn that for all the ways water benefits life, it also does life harm by degrading proteins and actually damaging DNA. (National Research Council, *Limits of Organic Life*, 27)

peroxide—can do the same job at least as well. Ammonia, for instance, can dissolve most organic compounds as effectively as or more effectively than water can, and it can also dissolve metals like sodium and magnesium directly into solution. If the life around us used liquid ammonia or hydrogen peroxide as its medium, we'd probably be marveling at how well its characteristics were suited for life's needs, until we realized that we had it backward—that life evolved to take advantage of those very characteristics.

In addition to an energy source to drive its chemistry and a medium in which that chemistry can work, life needs the chemicals themselves—that is, molecules that perform specialized functions that allow an organism to metabolize and reproduce. Familiar life maintains much of its metabolism with proteins—quite a lot of them, in fact. A typical cell contains 100 million proteins[3]—and most are specialized to perform specific functions. Some take energy from nutrients, others assemble more proteins, and still others dispose of waste and repel intruders.

Each of these functions is a complex process involving many steps. To take one example, when a cell somewhere in your body needs insulin, certain proteins inside the cell pull apart a section of the DNA molecule pairs, exposing the particular sequence of base pairs that signifies one of the many amino acids needed to make an insulin molecule. Other proteins read the sequence and make an ad hoc and temporary copy called messenger RNA. Then, still other proteins work over the messenger RNA, slicing and splicing until they've fashioned the amino acid needed. Finally, the molecules of protein and DNA called ribosomes (the structures that, you may recall, may set the lower size limit for a cell) pull the newly formed amino acid together with others made the same way by other proteins, and coordinate with other ribosomes, all now pulling and pushing their own amino acids, to assemble a molecule of insulin.

It's worth noting that the common metaphor for the activity of a living cell—a time-lapse video of a day of city traffic compressed into a few frenetic minutes—hardly does justice to its speed. In an average human cell, 2,000 new proteins are created each and every *second*. Belgian biochemist Christian de Duve observes that the speed of metabolic processes within the cell is "beyond the powers of our imagination."[4]

As complex as chores necessary to maintaining a metabolism are, they are in some ways mere prelude and preparation for the main event: reproduction. Familiar life can reproduce, of course, because cells divide. For cells with nuclei, it all begins inside the nucleus, when proteins don't pull apart merely a section of the DNA molecule; they unwind and unzip the entire molecule along its entire 3-billion-base-pair length. Other proteins then scan one strand, make a copy, correct and repair proofreading errors, and, from material in the surrounding cytoplasm, fashion a matching strand that winds together with the copy, base locking neatly to base. Then the parent DNA, its own strands zipped up and rewound, is pulled to one side of the nucleus, the child DNA is pulled to the other, and the nucleus itself is squeezed in the middle until it splits into halves. Shortly thereafter the cell does likewise, with each half holding a nucleus. Where there was one cell, now there are two.

If the metabolic and reproductive processes in a cell are precise and fast, they are also delicate. In order to work, they need a barrier that protects and insulates them from the harsh world outside, even as they take energy and nutrients from that world and discharge waste into it. For this reason, life based in chemistry may have another requirement: a semipermeable barrier. In familiar life, that barrier is the cell membrane, a structure composed largely of lipids, the large molecules that water will not dissolve.[5]

In a much-cited 2001 article, microbiologist Norman Pace made a case that all life—including what we've been calling weird—would use the same or similar molecules to metabolize and reproduce, going so far as to argue that even the sugars used by nucleic acids, given their evolutionary advantages, are probably universal."[6] While some astrobiologists think that Pace may overvalue the advantages of those particular nucleic acid sugars and undervalue life's ingenuity, most expect that life based in chemistry will use chemicals and processes that are roughly analogous to those used by familiar life. They also expect that if weird life doesn't use proteins and DNA per se, it will use molecules just as large. How large? A molecule of water has a molecular weight of 18. Some protein molecules have molecular weights in the millions. A DNA molecule from a multicellular organism is likely to have a molecular weight in the billions; such a molecule—untangled, uncoiled, and stretched out—would be a meter long.

Biochemists call these gargantuan molecular assemblies "macromolecules." These are not, we should note, haphazard constructions, and you can't make them with just any element. Macromolecules need "backbones"—that is, long chains or sequences of the same kind of atom. Carbon, of course, serves as the molecular backbone of all familiar life. Only one other element is capable of making such backbones, and that element is silicon.

SILICON LIFE

Ideas of silicon-based life have a surprisingly long history. Among the first hypothesizers was Herbert George Wells, a figure better known today for his science fiction than his science. But Wells did have scientific training, having studied at the Normal School of Science in London under the tutelage of Darwin's advocate T. H. Huxley. Wells was captivated by Darwin's ideas, and he reasoned that natural selection and competition would operate for living beings anywhere. Accordingly, the Martians of his 1897 novella *War of the Worlds*, having evolved in a world with weaker gravity, needed prosthetics to move about on Earth. Because Wells's Mars was Earthlike, its inhabitants had evolved with the same needs as Earth's organisms (water among them) and with the same frailties (a vulnerability to certain bacteria).

However, Wells also imagined organisms far weirder, and he imagined them in the context not of science fiction, but of science. Taking inspiration from an address by an English chemist named Emerson Reynolds, in 1884 Wells penned a short piece making a case for life based on silicon. Momentarily, he allowed his imagination free reign, writing of "visions of silicon-aluminium organisms—why not silicon-aluminium men at once?—wandering through an atmosphere of gaseous sulphur, let us say, by the shores of a sea of liquid iron some thousand degrees or so above the temperature of a blast furnace."[7] He acknowledged that such an idea was fantasy—"merely a dream" were his words—yet he allowed that a biochemistry based in silicon and aluminum, what he called "an analogue to protoplasm,"* was possible.

---

* The *protoplasm* is the fluid content of the cell, composed mainly of nucleic acids, proteins, lipids, carbohydrates, and inorganic salts. In the nineteenth

Since Wells, many science fiction authors have imagined silicon-based organisms. So have a few biologists—but only a few. Until recently, most suspected that the resemblances between carbon and silicon were superficial at best, and that any putative silicon-based biochemistry would meet with at least three showstoppers.

The first was silicon's relatively discriminating nature. Silicon was long thought to form stable bonds with only a handful of elements. The price paid for such exclusivity was that even silicon's macromolecules—those built on backbones of silicon atoms being "silanes," and those built on backbones of silicon atoms alternating with oxygen atoms being "silones"—were little more than endlessly repeating sequences of the same atoms, sequences that chemists, sounding rather like jaded music critics, call "monotonous." If nature playing on a theme of carbon could compose symphonies, it seemed that with silicon she could manage only hour-long compositions on two or three notes.

The second showstopper was that, unlike carbon, silicon was generally thought unable to form double and triple bonds, and so unable to capture and channel electrical energy.

The third showstopper was that unlike carbon-based compounds, silicon-based compounds are highly reactive with many naturally occurring chemicals on Earth, including water and oxygen. Drop some carbon tetrachloride into a beaker of water, and not much happens. It will sit there and stay stable, quite literally, for years. Do the same with its cousin silicon tetrachloride, and the compound will dissolve in seconds. Or, take the simple compound of carbon and hydrogen known as methane

---

and early twentieth centuries (and at the time of Wells's writing), the term referred to a substance that endowed the cell with the ability of self-replication.

gas, expose it to air, and you may expect them to coexist peacefully. But substitute methane's silicon-based counterpart silane, and you'd better put on your lab goggles first, because you are guaranteed the spectacle of spontaneous combustion.

Many biologists long believed there was a fourth reason to doubt the utility of silicon for life, and it was more or less self-evident. After oxygen, silicon is the second most abundant element on Earth—a fact made evident by a glance at a map of northern Africa, western China, or the western United States. Much of Earth's silicon, combined with oxygen, is silica—or sand. Clearly, life has all the silicon it could ask for. But with the exception of diatoms that use it in their hard cell walls and a few plants that use it in various supporting structures, familiar life has opted for carbon instead.

For decades, the case for carbon as the only viable basis for all living organisms seemed strong. Even Carl Sagan, who regularly inveighed against parochial thinking in the search for extraterrestrial life, had difficulty imagining alternatives. "When I try to think of other elements as a basis for life," he said, "I always wind up being what I call a carbon chauvinist."[8] Exactly how carbon became a cause for chauvinism is an interesting question. If we expect to answer it, we'll need to back up a bit.

THE STUDY OF CARBON

The periodic table, its 118 elements named and numbered, arranged by period and group, seems the very model of order—a cabinet exquisitely crafted and custom-made to contain all of nature's essences. But any suggestion of real containment is belied by the possibilities inherent in *combinations* of those essences. It is those combinations, and the temperatures and pressures at which those combinations occur, that make possible ships and shoes and sealing wax, and at a somewhat higher level of complexity, cabbages and kings—as well as every material thing there is, was, will be, and can be.

Exactly how do chemists know what a given combination will produce? The surprising truth is they don't—at least not always. Frank DiSalvo, a professor of chemistry at Cornell University, admits, "Much of what we come up with we happen on by trial and error, and we can't predict what we'll get ahead of time."[9] Even in an age of computer-generated 3D simulations of tumbling, pirouetting macromolecules, chemistry defiantly remains a science of measuring, mixing, and waiting around to see what happens—in other words, a science of experimentation.

Naturally, there has been more experimentation with some elements than others, and so much with carbon that it is accorded its very own specialization: the field of organic chemistry. That there are more organic chemists than any other kind is certainly owed to the field's commercial profitability. Organic chemists working in research divisions of pharmaceutical, polymer, and petrochemical firms have derived millions of compounds—and generated revenues that rival the GNP of many nations. But the number of organic chemists is also owed to chemists' quite natural interest in the chemicals that chemists themselves are made of, and to the endlessly fascinating nature of carbon.

And so comes a question. Is carbon the spoiled child of the periodic table? Might its astonishing chemical virtuosity, while much deserving of attention, have distracted scientists from seeing or seeking the innate talents of other elements—especially, considering our interests here, those conducive to biochemistry? Many biologists think not. They suspect carbon is essential to all life that is based in chemistry. Norman Pace, in the spirit of scientific humility, says, "I would never say never," but he nonetheless admits, "I'm not at all optimistic about finding non-carbon life."[10]

There are, though, a few dissenters. One is British biochemist William Bains, currently affiliated with the University of Cambridge Institute of Biotechnology. Bains looks younger than his fifty-odd years. With his pale complexion and rimless eyeglasses, you might guess him once upon a time to have captained his school's chess team, and to have been the sort of student who openly challenged his teachers. In fact, Bains *is* something of a provocateur. And he can make a case for silicon.

THE CASE FOR SILICON

Bains can answer the arguments against silicon point by point. He admits that some silanes are indeed "monotonous" as charged, but he notes that the right kind of monotony can work wonders of complexity and variation. After all, the 3 billion base pairs of the DNA molecule itself, that very template of all genetic diversity, are built from only four nucleic acids. Moreover, Bains says, the claim that silicon can bind with only a few elements amounts to "a myth."[11] It is held as a truth by many astrobiologists only because they are unaware of recent work in chemical engineering, work showing that under certain circumstances—most notably very cold temperatures—silicon can bond stably with many elements.[12]

The stability of any given chemical, of course, is directly tied to its temperature. For a living organism to continue to be living, its biochemistry must be stable enough that its cellular structures hold together, but not so stable that nothing moves—the latter condition being a fair description of death. A working biochemistry is a bit like a juggler. A juggler can't be too stable; he has to juggle, after all. Neither can he be unstable; that is, he can't drop a bowling pin or let one go flying off in the direction of the audience. To keep his audience interested, he must always seem about to miss a catch, without ever actually missing a catch. He must be continually *approaching* instability, without ever actually getting there. It is likewise with an organism.

The organic chemistry of familiar life is stable, but approaching instability within the temperature range at which water is liquid, and with extremophile modifications, somewhat beyond that range. The reason silicon has difficulty forming large molecules is that within that same range, its bonds are weak. But at much colder temperatures, where carbon-based chemical activity all but ceases, silicon chemistry is still active, and silicon's weaker bonds are strong enough to form long molecular chains. (Wells, by the way, had probably erred in placing his silicon-based organisms in warm environments.) Just as impressive, silicon can bond with those elements in ways that produce a variety of molecular structures, including rings and cages.[13]

Bains notes that because chemists have not made certain compounds with silicon doesn't mean they *can't* make them, and doesn't mean that, somewhere, nature hasn't made them already. "There is," Bains says, "no reason why a high-molecular-weight silane or siloxane [a compound of silicon, oxygen, and hydrogen or hydrocarbons] should not have a highly diverse, highly structured set of side chains, analogous to proteins, nucleic acids or carbohydrates."[14]

As to the charge that silicon cannot capture and distribute electricity, Bains summons a robust counterexample in certain polysilanes—that is, polymers with silicon backbones—that are semiconductors, and argues that other silicon-based compounds might channel energy as effectively. Bains contends that if biochemists and biophysicists suspect that silicon chemistry cannot channel energy, it is only because they assume it would use the same chemical pathways that organic chemistry uses. Silicon chemistry, he maintains, would have to use other pathways, and those pathways are yet to be discovered.

That no silicon life is evident on Earth is a point Bains does not dispute. In our oxygen-rich atmosphere, he notes, silicon biochemistry hasn't a chance. The problem—if it is that—is that silicon bonds with oxygen easily and readily, and the result is silicon dioxide, or silica. The bond between silicon and oxygen is strong and difficult to break. It is the reason that most of the silicon on (and in) Earth is locked into rocks and, like some imprisoned fairy-tale princess, made unavailable for other chemical and biochemical activities. Bains's point is that the apparent absence of silicon life on Earth is at best a reason to admit that it is unlikely here. It is not a reason to think it unlikely elsewhere.[*]

WHAT LIFE NEEDS (REDUX) AND WHERE TO LOOK FOR IT

Let me summarize. Probably all biologists agree that life based in chemistry needs a source of energy. Most believe that it needs a semipermeable barrier that does the work a cell membrane does for familiar life. Most also believe it needs macromolecules with

---

[*] Using radio and infrared astronomy, however, scientists have found that silicon is rare in deep space, especially by comparison with oxygen, carbon, and nitrogen.

backbones of carbon. A few biologists—Bains among them—believe that those macromolecules may also have backbones of silicon.

If we use this near consensus to guide a wider search for life, we can begin to check a few places off—but, as it happens, only a few. Most places in the known universe have available to them a source of energy, and with a few exceptions—like the inert interiors of certain planets and moons—most places are in thermodynamic disequilibrium. Complex chemistry, the kind necessary to making semipermeable barriers and macromolecules, seems present at many places in the Solar System, and there is no obvious reason not to expect it elsewhere.

As discussed earlier, most biologists believe that chemically based life needs a liquid medium. It is here that we meet real limitations that allow us to narrow the search criteria. The only places we are likely to find liquids of any kind—water, ammonia, methane, or something else—are on the surfaces of planets and moons having atmospheres and orbiting in thin shell-shaped regions around stars, or in the interiors of planets, moons, and smaller bodies heated by some other means. While some liquids, like liquid hydrogen, are for various reasons unlikely, others, like ammonia, are known to exist in abundance.

Bains has graphed the distances from the Sun at which elements and compounds that could support complex chemistry are liquid. In so doing, he has shown us how to keep the search selective; at the same time, he has defined many habitable zones for weird life. It is easy enough for us to overlay these zones onto the orbits of the Solar System's planets. Recall the placement of the habitable zone for familiar life, in the traditional version—from just inside Earth's orbit to just beyond Mars's. And recall the new version, which includes the traditional but also adds smaller zones beneath the surfaces of several moons of the outer planets. To astrobiologists grown used to these, the effect of the Bains overlay is agreeably unsettling. It is a bit like seeing a world map inverted or a chessboard turned 180 degrees halfway through a game.

Imagine the orbits of the planets of the inner Solar System—Mercury, Venus, Earth, and Mars—as four concentric circles on a dinner plate, with the Sun at the plate's center and Mars's orbit just inside the plate's rim. A wide ring representing the traditional habitable zone for familiar life covers the orbits of Earth and Mars. It goes dark, and at the orbit of Mars appears another ring—this representing the range at which, on a planet or moon with a substantial atmosphere, the Sun's warmth is sufficient to melt frozen hydrogen peroxide, but not so great as to boil it. A second ring, overlapping yet extending outside this ring, represents the range at which hydrogen peroxide could exist as a liquid beneath the surface of a planet or moon.

Zoom out for a field of view wide enough to accommodate the vastly larger scale of the outer Solar System, and the inner Solar System is shrunk to the size of a dime, now sitting in the center of the plate. Between the edge of the dime and the plate's rim appear double rings representing analogous ranges for other liquids. From innermost to outermost are double rings for ammonia, methane, ethane, and, near the rim, liquid nitrogen.*

---

* Recall that the traditional habitable zone around a star is termed, colloquially, the "Goldilocks zone." Presumably, the fair-haired heroine of the children's story was made of DNA, amino acids, and proteins. Suppose that somewhere out there in a small moon's subsurface ocean, or in the clouds of a giant planet, there are *other* Goldilockses—or at least microbial versions—and they are made of other stuff entirely. The zones they find habitable might be greater than ours by several orders of magnitude. Even within our local piece of real

All these liquids are "cryogenic," this being the term applied generally to things with low temperatures, and each has been shown to support complex chemistry. What of *biochemistry*—the chemistry, vastly more complex, that works in living organisms? Enzymes, which we might think of as molecules edging toward self-organization and replication, are known to work in solutions of methionine and ethylene glycol chilled to $-100°C$. Admittedly, it is a long way from such molecules to the breathtakingly complex structures and chemical systems that make a living cell, and it may be that at cryogenic temperatures such complexity is simply impossible. But a handful of scientists suspect otherwise. Inspired by the examples provided by extremophiles, they have imagined exotic biochemistries, including chemical pathways and catalysts, that would work in the very cold. They have also imagined organisms—some simple, some not so simple—that might use those pathways and catalysts. Without intending to, these scientists have given us a bestiary that lies somewhere between the known and the mythical, somewhere between the *Encyclopedia of Life* and Margaret Robinson's *Fictitious Beasts*. In other words, a bestiary of weird life.

---

estate—the Solar System—there are more than 160 moons orbiting the outer planets, four dwarf planets, and billions of comets. If you believe most who tell the story of Goldilocks, its lesson is to respect the possessions and privacy of others. There is, though, a deeper truth in the story; it is that despite what Goldilocks thinks, the world is not made for her. If advocates of the search for weird life are right, then we will learn, once again and in ways we had not imagined, that it is not made for us either.

CHAPTER FIVE

A Bestiary of Weird Life

Thermometers in the inner regions of Antarctica typically drop to −70°C in winter, and the official record for the coldest temperature on Earth, measured at the Russian research facility Vostok Station, is −89°C. Strange to say, such temperatures would not be unusual for a summer morning on Mars's Chryse plain, the landing site of *Viking 1*—a fact that, depending on your point of view, makes Antarctica seem very far away or Mars seem not so distant. Of course, there are much colder places in the Solar System, but if we wish to visit them we must leave Earth and Mars far behind, and venture outward nearly 800 million kilometers to the vicinity of Jupiter, where the Sun seems pale and shrunken. For the moment let's turn our attention not to Jupiter itself, but to three of the planet's so-called Galilean moons—Europa, Ganymede, and Callisto. Each is large enough to be called a world.

It's cold here. Page through polar explorers' descriptions of Arctic winters or jilted lovers' descriptions of hearts gone cold, and you won't find words adequate to describe the chill. The surface temperature of Europa, at its equator, averages −160°C. Its smooth surface is covered with great sheets of fractured ice, making for a crust that may be tens of meters or tens of kilometers thick, but beneath which almost certainly lies a vast, dark ocean. Many scientists believe there may be life in that ocean, taking warmth and energy from hydrothermal vents.*

In the view of many, Europa has overtaken Mars as the place in the Solar System most likely to harbor extraterrestrial life. In fact, although a Europan ecology is decidedly hypothetical, NASA scientists worried that *Galileo* might someday impact Europa and contaminate it. So in September 2003, when the *Galileo* orbiter had completed its eight-year reconnaissance of Jupiter and its moons, they steered the spacecraft into Jupiter's atmosphere at a speed of 174,000 kilometers per hour, thus ensuring that it, and any terrestrial microbes it might have carried, would be incinerated.

Well before *Galileo* met its timely end, there had been numerous proposals for unmanned missions with a focus on Europa: one for a spacecraft that would orbit the moon and map its subsurface with ice-penetrating radar; another for a spacecraft that would launch a projectile to impact the crust and generate a plume that the spacecraft would fly through and analyze; yet another—the most ambitious—for a lander that would drill and or melt its way through the ice and into the dark ocean beneath,

---

* A possible model for that ocean is Antarctica's Lake Vostok, a body of liquid water the size of Lake Ontario, lying beneath more than 3 kilometers of ice. It may be as much as a million years old, and its waters may have been separated from the rest of Earth's water for far longer. As this book goes to press, Russian scientists (following considerable controversy arising from concerns that they might unintentionally contaminate the lake) have obtained a water sample.

where it would release an autonomous submersible that would cruise those depths seeking life. At present, study teams are preparing a proposal for a scaled-back Europa orbiter that might stay within NASA's recently diminished budget, and NASA's European counterpart, ESA, is proceeding with its own plans for an independent mission.[1]

In the meantime, astrobiologists are making do with hypotheses of Europan life. The difficulty here is that those hypotheses depend on unknowns like the thickness of the ice crust, the depth of the putative ocean, the energy sources available, and, of course, the chemistry of the water. Some have suggested a water-ammonia mix.[2] Enough ammonia—say 30 or 40 percent—would mean a pH high enough to denature the DNA of any familiar life, and William Bains makes a case that in such an environment no terrestrial cell would have the energy for the chemical reactions necessary to maintain a balanced pH inside itself. Life in such an ocean would need an "ammoniated" biochemistry, making it unlike any life we know.[3]

Others, however, suspect that Europan life might greatly resemble familiar life. *Galileo*'s mission scientists found that Jupiter's magnetic field fluctuates around Europa, thus suggesting that water in its ocean is highly conductive. Kevin Hand, a planetary scientist at NASA's Jet Propulsion Laboratory, is one of many who suspect that the ocean is saturated with salts, mostly magnesium sulfate. "There are terrestrial halophiles, salt-loving microbes," says Hand, "that could survive in the ocean we propose."[4] In fact, it is possible that terrestrial halophiles are there already. In a series of computer simulations, a team of researchers from the National Autonomous University of Mexico simulated meteor impacts on Earth and found that rock fragments launched with sufficient speed by such impacts could enter orbits that eventually brought them to Jupiter—not quite Europa, but very near it.[5] As noted several chapters back, certain micro-organisms are capable of surviving such journeys. This means that there is a possibility—it should be emphasized a very remote one—that Europan life does not merely resemble familiar life. It *is* familiar life.

Europa and her sister moons are each several thousand kilometers in diameter—the size of small planets. Bodies much smaller than this, less affected by tidal heating, were once thought unlikely to hold liquid water. One such is a satellite of Saturn called Enceladus, a small snowball of a moon (with a diameter of 500 kilometers) that might have come straight out of *The Little Prince*. When in 2005 the *Cassini* spacecraft returned images of the moon, mission scientists were properly stunned to see it pinwheeling geysers of water ice and ammonia crystals into the vacuum—by what internal forces they could only guess. Since the discovery, planetary scientists have suspected that Enceladus does experience tidal heating after all, and that the geysers arise from long-lasting fissures and cavities holding liquid water along with organic molecules, nitrogen, and mineral salts. Such places might be agreeable abodes for life, and as with possible life in Europa's ocean, it might resemble certain Earthly extremophiles.

Enceladus may have more accommodations for life than fissures and cavities. In 2011 John Spencer, a *Cassini* mission scientist, told a journalist, "Basically, I suspect we have an ocean."[6] To be able to say things like "basically, I suspect we have an ocean" as part of your day job, you might think that 1.2 billion kilometers is not too far to send a spacecraft. But you might also think it seems a long way to go to find microbes like those we know, and certain astrobiologists would agree. Europa and Enceladus are no doubt intriguing. But if we are inclined to look for rather more exotic life—life that, say, could use liquid methane as its solvent—we might look elsewhere, and at one place in particular.

A WORLD FOR WEIRD LIFE

In the seventeenth century, a Dutch natural philosopher named Christiaan Huygens contributed to studies of motion and gravity, proposed a wave theory of light, and invented the pendulum clock. He also designed several new sorts of telescopes, and spent enough time looking through them to earn a reputation as one of the best observational astronomers of his age. Although Huygens took care to avoid mixing seeing and imagining (dismissing Johannes Kepler's claim to have observed artificial constructions on the Moon as a "pretty story"), he was open-minded on the matter of life elsewhere. In a 1698 work elegantly entitled *Cosmotheoros, or, Conjectures concerning the Celestial Earths and Their Adornments*, Huygens asserted that given one example of a life-bearing world, we have no reason to conclude that others are barren. In fact, he made a caveat that might allow us to claim him as an early proponent of weird life. He maintained that given the Moon's evident lack of atmosphere and water, life there would necessarily be quite different from that we know.

In a 1656 treatise, Huygens announced his discovery of a moon orbiting the planet Saturn. Three years later he published *Systema Saturnium*, in which he described the moon's period of revolution as slightly less than sixteen days, very near the value derived by later observations. Huygens began to call it "my moon," and other astronomers obligingly followed suit, calling it "Huygens's moon." The practice continued until the late nineteenth century, when, following the suggestion of English astronomer William Herschel, most astronomers began to call it Titan.

Titan was large for a moon, roughly the size of the planet Mercury. For the next three centuries little else was learned about it. Even in the most powerful telescopes, astronomers could see only a featureless reddish orb, its surface hidden beneath a dense, opaque atmosphere. In the 1950s astronomers detected methane in that atmosphere; some thought it evidence of an ocean of hydrocarbons. In the 1990s, radar signals bounced off Titan's surface suggested land terrain in some parts, seas or large lakes in others. In the mid-1990s the Hubble Space Telescope imaged the ground in near-infrared light and revealed light and dark features. But exactly what those features represented was an open question, and Titan's surface remained a mystery.

In 1997, NASA, in conjunction with the European Space Agency (ESA) and the Italian Space Agency, launched the probe *Cassini* (named after the discoverer of Saturn's rings), carrying a Titan lander named *Huygens*, a spacecraft that, for all its technical sophistication, had an appearance that was particularly unprepossessing, looking like a large, inverted pie tin. In fact, the design was a frank admission by the engineers that their reach had exceeded their grasp. Because no one had ever seen the surface of Titan, no one knew whether the lander would come to rest on solid—that is to say, frozen—ground, or in a lake or sea, or in a methane slush. So *Huygens* was designed to withstand a hard impact on rock or ice. It was also designed to float.

On Christmas Eve, 2004, *Cassini* was in orbit around Saturn and it released *Huygens*. Three weeks later, the lander fired its small vernier engine and began a long arcing fall. For an hour and forty minutes, its camera made 3,500 images. They have since been assembled into mosaics and sequenced into a video that represents the view from the lander through its entire descent, making for a record as dramatic as any in the history of human exploration. For anyone who cares to look, the video is posted online, accompanied by the sound of pinging instruments.* It begins as *Huygens* nears the fringes of Titan's atmosphere, and the moon appears as a reddish sphere, with no distinguishable features. Soon the lander ejects its heat shield and deploys a parachute, slowing its descent. Forty minutes in, Titan seems to develop dark stains over large areas. The rate of exposures increases, the pings come faster, and the stains resolve themselves into badlands, rough hills, and arroyos that might be part of a mountainous desert on Earth. What seems a canyon appears in the left of the frame, and for a moment we see rugged mountain ranges like outspread fingers, cut through with complex networks of channels. Then, abruptly, we are seeing a still image: roundish, dust-covered ice pebbles in the foreground, giving way to a flat plain that stretches to the horizon—all beneath a hazy reddish sky.

In the hour and ten minutes available before its batteries died, *Huygens* performed a preliminary reconnaissance. Its instruments detected methane, presumably sublimed into the local atmosphere when the lander heated the ice beneath it. Meanwhile, *Cassini* began to map the moon's surface with both radar and near-infrared imaging, and it would continue to do so for years. Much of that surface was water ice, as hard as granite. Near the equator were fields of dunes made not of silica, but of a mysterious organic material having the texture of coffee grinds. The dune fields stretched for hundreds of kilometers, and radar

---

* http://www.nasa.gov/mission_pages/cassini/multimedia/pia08117.html.

showed some dunes to be 150 meters high. Elsewhere were great shield volcanoes built by discharges of "cryomagma," a slushy mix of melted water and ammonia pushed upward from a sub-surface ocean 480 kilometers below the ice crust. In the far south, river channels had carved great canyon systems; and in the northern latitudes were more channels—some dry, some perhaps not. In the same latitudes were lakes of liquid methane and ethane—one as large as Lake Ontario. These were—and are—the only bodies of liquid known to exist on the surface of any world other than Earth.

It is precisely because liquids have shaped and still shape its surface that, despite the frigid temperatures, Titan is the most Earthlike of any planet or moon in the Solar System. If we could enter a Titanian landscape and, say, walk along the shore of a northern lake, we would find the scene at once strange and familiar. We would see foaming waves lapping gently at a pebbled shore, and that shore curving around a large bay framed on either side by low, weathered hills—all lying uneasily under a reddish twilit sky. We might guess we were on a particularly inhospitable and rocky coast somewhere on Earth, following an evening storm. If we stayed very long, of course, we would learn that the daylight sky always seems twilit, and if we looked closer at the beach we would find that the pebbles were water ice and the surf was methane and ethane.

LIFE ON TITAN

The authors of the NRC report were particularly intrigued by Titan. Its atmosphere was in thermodynamic disequilibrium. Its surface was cold by Earth or Mars standards but not too cold for chemical bonds; in fact, ground near *Huygens* had many molecular compounds that contain carbon. And the surface had not one but *two* mediums and solvents in which life might arise: liquid methane and a slush of water and ammonia. All these findings led the report's authors to realize that Titan might help them understand the relation of life to chemistry. Envisioning Titan as a "control" in what might well be the most profound life sciences test possible, the authors of the NRC report put forth a startlingly simple hypothesis—one whose resolution would explain that relation decisively. "If life is an intrinsic property of chemical reactivity," they wrote, "life should exist on Titan."[7] But what sort of life?

A year before *Cassini* entered Saturn's orbit, Steve Benner and several colleagues suggested that on Titan, liquid hydrocarbons like methane might fill the role of *biosolvent*, a liquid medium that allows and facilitates chemical reactions conducive to life.[8] Soon after, two teams of scientists took up the idea. One team was made up of Chris McKay of NASA's Ames Research Center and Heather Smith of the International Space University in Strasbourg, France. The other team consisted of astrobiologists Dirk Schulze-Makuch (who, recall, had hypothesized hydrogen peroxide–drinking Martians) and David Grinspoon. Both teams calculated the energy available for methanogenic life in Titan's atmosphere, and both teams drew the same conclusion—that on Titan, hydrogen might play the role that oxygen plays in the biology of Earth.[9] Organisms on Earth metabolize with energy derived from the chemical reaction of oxygen and organic material and produce carbon dioxide and water as waste. Titanian organisms, so these scientists thought, might metabolize with energy derived from the chemical reaction of hydrogen and organic material, and produce methane as waste.

McKay and Smith recommended that their colleagues, by

way of testing the hypothesis, look for chemical disequilibrium. It was exactly the sort of test that James Lovelock had advocated to detect life on Mars some half a century earlier. If Titan's atmosphere was in equilibrium, scientists should expect to find significant amounts of certain chemicals—three in particular. There should be quite a lot of ethane, enough to submerge the moon's entire surface to a depth of several meters. There should be enough acetylene (produced through reactions triggered by ultraviolet sunlight) to be detected by *Cassini*'s instruments. And there should be a good deal of hydrogen—produced constantly through reactions triggered by ultraviolet light, with some rising through the atmosphere and leaking into space, and some sinking toward the surface. McKay and Smith posited that if *Cassini/Huygens* mission scientists found an *absence* of these chemicals—or in the words of the paper, "anomalous depletions of acetylene and ethane as well as hydrogen at the surface"[10]— they might also have found evidence of life. It would, of course, be a very specific sort of life—organisms that would absorb ethane, acetylene, and hydrogen, and produce methane as waste.

Over the next several years, data from *Cassini* and *Huygens* were analyzed. In several attempts to identify ethane and acetylene, the mission scientists found none of either. In 2010, a computer simulation of Titan's atmosphere suggested that hydrogen was not accumulating near the surface, and that there was the strong possibility of a flux of hydrogen *into* that surface.[11] All evidence implied that Titan's atmosphere was in disequilibrium. Of course, the simulation of disappearing hydrogen might be wrong, and in an interview with *New Scientist*, McKay was properly cautious, allowing that what they had was at least a "very unusual and unexplained chemistry."[12]

Lovelock had said that he did not expect life to survive in

pockets on otherwise barren bodies, except for the brief moments when it was gaining a foothold or just before being extinguished completely. For by far the greater part of its existence, he argued, life on a planet or moon would be widespread. Therefore, we would expect to find only two categories of extraterrestrial worlds: those that were barren and those that were rich with life. Earth, of course, is in the latter category; as we've noted, life is very nearly everywhere there is liquid water. McKay and Smith suggested an analogy on Titan, with the role of water played by methane. Since the chemical is everywhere on Titan—the landing site of *Huygens* was moist with methane, its atmosphere has methane-nitrogen clouds, methane rains down on its surface, and the northern hemisphere is dappled with methane lakes—they predicted that if there were *any* methanogens, there would be a great many, and they would be widespread.

Cassini has since completed its initial four-year mission, as well as an "extended" mission. It is now in its second extended mission to explore the Saturnian system, expected to last through September 2017.* At present there are ideas for exploring Titan remotely, using some unusual and original designs. One idea is for a mission that would be more or less a repeat of *Huygens*, but targeted at one of the northern lakes or a shoreline. Titan, though, is a world that invites exploration with vehicles utterly unlike the rovers of yore. Another idea is for an orbiting space-

---

* In February 2009, NASA and ESA officials agreed to continue pursuing studies of a mission to Jupiter and its four largest moons, and to plan for a mission to Titan. NASA and ESA agreed that the Jupiter mission was the most technically feasible to do first, but ESA's Solar System Working Group recommended, and NASA agreed, that both missions merited implementation. By early 2012, however, budget cuts forced NASA to suspend plans for both missions. ESA is proceeding on its own, developing a version of the Jupiter mission.

craft, teamed with a surface probe and—most ambitious—a balloon equipped with a helicopter rotor that would fly wherever the mission scientists directed it.[13] Those who imagine exploring Titan are especially intrigued by its lakes. Still another idea is for a boat—specifically, a saucer-shaped probe that might parachute and splash down in one of those lakes, where it would drift with wind and currents, taking measurements of temperature and methane humidity, as well as images of the shoreline.[14]

HOT JUPITERS AND OCEAN PLANETS

In 1992, astronomers confirmed the first discovery of two planet-sized bodies outside our Solar System. They were orbiting the remnant of a star that had exploded as a supernova, leaving a compact sphere of neutrons spinning furiously at thousands of times a second, and sending pulses of electromagnetic radiation streaming into space. No one expected planets to have survived the unimaginable violence of a supernova explosion, and no one knew—and in fact, no one knows still—exactly what those two bodies (and a third confirmed later) are. They may be the cores of extrasolar Jupiters with their atmospheres ripped away, the remnants of a companion star, or something else.

Soon enough, planetary scientists would identify stellar phenomena that were rather more conventional—but perhaps more exciting. In 1995, astronomers discovered evidence of a body orbiting a star like our Sun. It was a planet by anyone's definition, and it was found by identifying the tiny but measurable wobble a planet produces in its parent star as it orbits that star. The "wobble detection" method proved reliable and quickly led to the discovery of many more planets, but it had a built-in bias, as it was best suited to detecting large wobbles produced by large planets. These were enormous worlds that astronomers call "hot Jupiters"—"hot" because they are nearer their parent star than the planet Mercury is to our Sun, "Jupiters" because they are at least as massive as that world.

In the 1990s, as instruments and techniques were refined, astronomers found smaller planets in wider orbits, and the rate of discovery increased so quickly that by the first years of the twenty-first century, a new world was swimming into our ken every two weeks. There was a steady stream of planetary firsts: the first planet discovered orbiting two stars, the first orbiting a red dwarf star, the first Neptune-sized planet, and so on. As this book goes to press, scientists have identified nearly 800 planets and imaged several in the infrared. A handful of these planets have orbits that take them on a direct line of sight between their parent star and us, such that, in the field of view of a hypothetical telescope far more powerful than any in existence, the atmosphere of any of them would appear as a bright ring around an otherwise darkened sphere. Even with existing telescopes, astronomers have been able to conduct spectrographic analyses of such atmospheres, and have found them to contain (among other gases) sodium, water vapor, and methane.

Most planets, though, are not so cooperative as to pass between their star and astronomers or astronomers' instruments. To understand the chemistry of their atmospheres, or much of anything else about them, scientists call upon elaborate models of planetary formation. One such model implies that planets like Neptune, whose core is surrounded by a thick shell of water ice, can be pulled into orbits so near their parent star that the ice melts, making for planetwide oceans tens or hundreds of kilometers deep, giving way at the greatest depths not to a floor of sediment, basaltic rocks, and hardened magma as on Earth, but to a peculiar form of ice that forms under great pressures. Although such a planet, like most planets, would receive a continual peppering of meteorites and so a steady ration of metals and silicates, many planetary scientists suspect it would not be enough to sustain complex chemistry, let alone biochemistry. Those vast oceans would likely be pure—and sterile.

Other models of planetary formation imply more Earthlike worlds, which would have ample water near their surfaces or on them, as well as mantles of silicates and metals—ingredients necessary to any biochemistry. The search for such worlds is the purview of NASA's Kepler space observatory.

THE NEW WORLDS OF KEPLER

Planetary scientists now suspect that Earthlike planets may be quite common in our galaxy (although, according to one hypothesis, limited in distribution to a "galactic habitable zone"); in fact there is a part of space where a great many such worlds have already been identified. It is a patch of sky between the constellations Cygnus and Lyra, visible on a clear summer night from anywhere in the Northern Hemisphere. This is the star field being observed by Kepler, whose photometer can detect the slight darkening and brightening of a given star, the darkening and brightening that may mean a planet has passed in front of it—or, as an astronomer would say, made a "transit." Kepler's mission rules require three transits to confirm a discovery, and as of February 2012, Kepler's scientists had made provisional identifications of 2,321 new worlds, with forty-six orbiting inside their parent star's traditional habitable zone.

There are millions of stars in Kepler's field of view. Given practical limitations of time and budget, the mission scientists cannot look for transits of all of them, but they can look for about 145,000. For various reasons, they are concentrating their efforts on stars belonging to a class of which our Sun is a member, stars that burn more or less steadily for roughly 10 billion years. The selection pleases astrobiologists, who suspect that life requires a source of energy that is stable for long periods. Although the first life arose on Earth 3.5–3.8 billion years ago (almost as soon as it was possible), complex cells with nuclei have been here for only 2 billion years, and multicellular life—complex life—for only 1 billion. It seems that complex life on Earth needed nearly 3 billion years to establish itself, and, owing to a warming sun, it could not have taken much longer. The Sun's temperature is increasing such that in a billion years it will have become about 10 percent warmer than it is at present, boiling away Earth's oceans, baking its surface, and making for an environment in which life as we know it could not survive.[15]

Complex life on Earth, then, had a range of some 5 billion years in which to establish itself, and it hit a mark inside that range, with roughly 2 billion years to spare. But suppose, as theoretical physicist Brandon Carter suspects, that we are a special case, and that on average, complex life takes longer than 5 billion years to gain a foothold.[16] If Carter is right, planets orbiting stars like our Sun would be poor bets for life.

OTHER TITANS

Planetary scientist Jonathan Lunine observes that the smaller, cooler sorts of stars that astronomers call red dwarfs are also stable—much more so. Because their nuclear reactions are far slower than those of larger stars, red dwarf stars are expected to burn for *trillions* of years. Life on a planet orbiting a red dwarf,

Lunine notes, would have much more time to arise, and much more time to survive. Nonetheless, such life—if it were life like that we know—would face challenges. Red dwarfs burn at such low temperatures that a planet orbiting a red dwarf star, to be warm enough to have liquid water on its surface, would have to be ten times nearer that star than Earth is to the Sun. Such proximity would expose the planet to intense flares and stellar winds and, perhaps worse, ensure that it was "tidally locked," with the sunlit side forever baking, the darkened side eternally frozen.

It is true that liquid water might exist and life might survive in the twilit regions. It is also true that scientists have models suggesting that a sufficiently dense atmosphere might work to moderate temperatures, and that some of these planets might have such an atmosphere. But Lunine suspects that most are lifeless. He also believes, however, that inhospitable planets in Earthlike orbits are not a reason for astrobiologists to take a pass on red dwarf systems. If we want to find a planet with temperate conditions in such a system, he advises that we look to an orbit at about the same distance from the star as Earth is from the Sun—about 150 million kilometers out. From that distance the effect of most flares and stellar winds would be greatly diminished. Moreover, any planet orbiting at that distance would not be tidally locked, so temperatures across its surface would moderate.

Of course, any water on that surface would be frozen, as would any ammonia. In fact, such a planet would be as cold as Titan. But as we've seen, an environment like Titan's could be home to certain kinds of life. And because red dwarfs greatly outnumber Sun-like stars by a ratio at least ten to one and perhaps as much as a hundred to one, the galaxy may have far more Titans than it has Earths. If such worlds *are* congenial to life, then life like that hypothesized by McKay and Smith may be far more common than familiar life. In which case, statistically speaking anyway, life on Earth might be the life that's weird.

AND COLDER STILL

Suppose we return to our own Solar System, but venture now more than 4 billion kilometers out to the orbit of Neptune, the outermost of the planets, from where the Sun appears as only a bright star. Our interest is in Neptune's largest moon, Triton. The most detailed images we have of Triton were made by *Voyager 2* during its 1989 flyby, and since the moon's atmosphere is the thinnest wisp of nitrogen, they showed a clearly visible geography. There were craters, impact basins, and vast snow-covered plains. A third of the moon's surface was an area, long since melted and refrozen into thousands of circular indentations, that looked like nothing so much as the ice-rimed skin of a cantaloupe. There was some water ice, but most was of the type chemists might call "exotic"—carbon dioxide, methane, and nitrogen—and you and I would call very, very cold. A typical midday temperature on Triton might be –235°C, a number whose chill is better appreciated as 35 degrees above absolute zero, the temperature at which frigid turns decidedly rigid, and all molecular motion ceases. Yet evidence suggests that there's liquid there, and it is near the surface.

Voyager imaged several long, parallel dark streaks on Triton's surface. They are something of a mystery, and many planetary geologists suspect that they were produced by geysers. The thinking is as follows. Parts of the moon's surface are clear nitrogen ice; beneath it is more nitrogen ice mixed with organic material. In the weak light of the distant Sun, the clear ice on the surface acts like greenhouse glass, making for, one must note, a very chilly greenhouse: at a pressure of 1 atmosphere, nitrogen melts at 63 degrees above absolute zero and boils at 77 degrees above absolute zero. Nonetheless, the frozen nitrogen beneath the clear ice melts, gurgles, and trickles, and mixes with nearby organic material. Sooner or later the nitrogen boils and erupts violently through the surface ice, carrying the organics with it. In Triton's low gravity, the mixture gushes upward for several kilometers until it meets prevailing winds that carry the organics great distances before they fall to the surface. Where they fall, they leave the dark streaks.

Building on what little is known and can be surmised of Triton's geology and subsurface chemistry, William Bains has conjectured a way that a complex chemistry involving silanols and silanes, reacting in liquid nitrogen and driven by heat from Triton's core, might set the stage for a biochemistry—a system that organizes chemistry and, through feedback mechanisms, sustains itself.[17]

SINKERS, FLOATERS, AND HUNTERS

The mediums used by the hypothetical life we've discussed so far are liquids—one of the three states of matter that, recall, are termed "classical." There are several nonclassical states of matter, and a number of scientists have imagined organisms that might use them. To find their habitats, we'll need to reverse course and head Sunward, to the warmer regions of the inner Solar System. We begin with the planet Venus.

The Venusian surface bakes at blast-furnace temperatures in excess of 460°C, and beneath a dense atmosphere of carbon dioxide that weighs on it with a pressure ninety times that exerted by Earth's atmosphere at sea level. In the 1960s and 1970s, the Soviet Union parachuted ten probes to that surface. They were built like bathyspheres, but only a few survived more than an hour. It was hard to conceive of an organism that might do better. But some 40–70 kilometers above that surface, things are rather different. Venus's atmospheric pressure is only half again as great as Earth's at sea level, and temperatures average 37°C— what we might expect on a warm day in the tropics. At such altitudes there happen to be large volumes of droplets of liquid suspended in a gas—a nonclassical state of matter that chemists call *aerosols*, and you and I call clouds.

If you think ideas of organisms living in Venusian clouds might seem to have left reason somewhere in the lower troposphere, consider that the cumulus and stratus in our own atmosphere are well populated with bacteria, algae, and fungi. Bacteria survive in clouds for very long periods, and they can do so because there is water (of course), but also because most clouds hold significant quantities of nitrogen, sulfur, and various organic acids—chemicals a bacterium regards as food. It seems that quite a few are at dinner. A cloud sample taken by a French meteorological station was found to be home to no fewer than seventy-one bacterial strains, many individuals of which came from oceans, presumably pushed into the air by a breaking wave or bursting bubble and lofted further by wind.[18]

Microbes of various sorts can get higher still. In 1978, researchers using meteorological rockets fitted with samplers found bacteria in the mesosphere, that layer of atmosphere above the stratosphere, at altitudes of 50–100 kilometers. (By way of comparison, commercial aircraft fly at altitudes of 10 or 12 kilometers. The Fédération Aéronautique Internationale, the international standard-setting and record-keeping body for aeronautics and astronautics, puts the boundary between Earth's atmosphere and outer space at an altitude of 100 kilometers, about 62 miles.) At least as surprising as the microbes' sheer presence was the discovery that some, otherwise identical to their Earthbound brethren, had evolved the ability to synthesize pigments as a resistance to ultraviolet radiation.[19]

But back to Venus. Earth's clouds are essentially water droplets suspended in the gases of our atmosphere—mostly nitrogen and oxygen. Venusian clouds, on the other hand, are a fine mist of droplets of sulfuric acid suspended in carbon dioxide. Along with the temperatures and pressures, they would seem to complete a picture of Venus as hell—would, that is, if sulfuric acid were as harsh as its reputation. In fact, though, that reputation is only partly deserved. When the chemical is used as an industrial solvent, it is mixed with water, and then it's the water that does the corroding and dissolving; the sulfuric acid is only a catalyst. *Pure* sulfuric acid—the sort in Venus's clouds—is a mild solvent at best. And a mild solvent, just strong enough to pull apart some molecules and free up carbon for chemical reactions, is the kind a metabolism prefers.

The metabolisms of known acidophiles, by various mechanisms, keep water in and acid out. Astrobiologists Dirk Schulze-Makuch and Louis Irwin suggested that the metabolisms of Venusian cloud dwellers would need to reverse the exchange, keeping the acid in and the water out. Strange as such a metabolism may seem, scientists know of nothing to prohibit it. Certain plants are known to use acids to synthesize molecules,[20] and there are several hypotheses for whole metabolisms that might use sulfuric acid. In 2002, Schulze-Makuch and Irwin argued that the possibility of acidophilic, cloud-dwelling Venusians was sufficient to justify a mission that would scoop up a wisp of Venusian atmosphere and return it to Earth for study.[21]

Ideas of cloud-borne Venusians are not new. As far back as 1967, when there was evidence for substantial amounts of water vapor in the planet's clouds, Carl Sagan and Yale biophysicist Harold Morowitz hypothesized organisms the size of Ping-Pong balls, with skins a single molecule thick.[22] Such organisms, so their thinking went, might have originated on the surface sometime in a distant past when conditions there were more temperate and, as the surface heated up, migrated to the skies. Like jellyfish in terrestrial oceans, they would maintain buoyancy with float bladders—filled, in the Venusian case, with hydrogen. The idea was criticized because although Morowitz and Sagan had suggested ways the cloud-borne might be cloud-born (sexual and asexual reproduction), they had not shown how they might have evolved, and no one could imagine an evolutionary path by which a Venusian surface-dwelling organism might develop a float bladder. Sagan, though, was undeterred. He didn't abandon the cloud-borne life; he just suggested another place to look for it.

If you could peel Venus's atmosphere like the skin of an orange, stretch it out, and flatten it against the planet Jupiter, it would occupy only about the area, relatively speaking, that India occupies on the surface of Earth. The atmosphere of the larger planet is vast. It is also deep. In 1995, NASA's *Galileo* atmospheric probe fell nearly 5,000 kilometers through Jovian air, measuring the wind, temperature, composition, clouds, and radiation levels all the way, and ceasing to function only when it reached an altitude where the pressure was twenty times that of Earth's at sea level. But during that long fall it taught us a great deal.

Jupiter's upper layers are a clear thermosphere and stratosphere of hydrogen and helium. Beneath these is a troposphere of clouds and hazes of ammonia, ammonia hydrosulfide, and water—all smeared by the planet's swift rotation into bands whose edges are frayed into wisps the size of continents. Beneath that, things get very strange. Jupiter has no solid surface. Instead, somewhere beneath the troposphere, at a level far deeper than the *Galileo* probe penetrated, the atmospheric pressure increases to the point where the hydrogen and helium become a "supercritical fluid"—a nonclassical state of matter with the properties of both a liquid and a gas.

Twenty years earlier, scientists knew far less of Jupiter's chemistry and cloud systems, but they knew enough that Sagan and astrophysicist Edwin Saltpeter could hypothesize Jovian life in considerable detail. The result, a 1976 article published by the American Astronomical Society, counts as one of the few attempts outside science fiction to describe not merely weird organisms, but an entire weird ecology.[23]

In that article, Sagan and Saltpeter imagined cell-sized organisms they called "sinkers." They looked like tiny hydrogen-filled balloons, and they could drift in the upper troposphere for weeks or months before they fell to lower levels, where the high temperatures would be fatal. If their species were to survive longer than that, at least some would need to reproduce before they fell. Sinkers, so the paper continued, might reproduce asexually by exploding seeds or spores, or they might coalesce much as raindrops do, with others of their kind, thereby becoming a single, larger organism.

The authors knew that the charge against the hypothesis for the Venusian organisms might be used here as well, and they had prepared an answer. The hydrogen gas makes a sinker buoyant (slightly lighter than an equal volume of Jupiter's hydrogen-helium atmosphere), while the thin skin weighs it down. Since, as with coalescing bubbles, the volume-to-surface (or hydrogen-to-skin) ratio increases with every coalescence, enough coalescing will make a sinker unsinkable. It will become another organism, of the type Sagan and Saltpeter called "floaters." Naturally, floaters would mate with other floaters. Over time, genetic variation and natural selection would produce floaters with sense organs and floaters that could direct their flight. Nothing would limit their growth, and nothing would prevent them from reaching dimensions befitting the vastness of their habitat. Sagan conjectured that they might become "kilometers across, enormously larger than the greatest whale that ever was, beings the size of cities."[24]

Life in Jupiter's clouds would be anything but dull. Since pure hydrogen gas is all that would keep a floater from a fiery death, it would be regarded as a valuable commodity. And since it might be easier to steal hydrogen from a floater than to separate it out of the atmosphere, some floaters might evolve into a third type of Jovian: "hunters." The hunters, too, might grow to enormous sizes. Sagan and Saltpeter conjectured that all three types—sinkers, floaters, and hunters—might be stages in a single life cycle. In any case, there was the possibility of a dynamic and dramatic ecology.

Sagan had a particularly visual imagination. A few months before the paper on life and death in Jovian skies appeared, he had suggested that *Viking*'s scanners image Earth from the Martian surface (they were insufficiently sensitive). In 1989, he would propose that the *Voyager 2* spacecraft, at a distance of 4.8 billion kilometers from the Sun and leaving the Solar System, should image our home star's retinue of planets (it did). It is not surprising, then, that he noted that floaters and hunters, if they existed, would be of such size that they might be resolved by the imaging system that would be installed on the twin *Voyager* spacecraft. The *Voyager* mission planners made no special effort for that system to target such creatures, and alas, nothing in the *Voyager* images, or any images made of Jupiter since, might be mistaken for them.

CHAPTER SIX

Life from Comets, Life on Stars, and Life in the Very Far Future

The Space Science and Astrobiology Division at NASA Ames is housed in a concrete-and-glass structure of 1960s vintage, and you might mistake it for a classroom building at a community college, until you enter the lobby. On the far wall is a large mural depicting an imaginative history of life in the universe. From left to right, star clusters are braided into DNA molecules that morph into cells, leading across billions of years and ever-larger plants and animals, to finish with a pod of blue whales cavorting in water shot through with sunbeams. It's a vision breathtaking in its richness and grandeur, and it bespeaks a seriousness of purpose in the people who work here.

The vision is offset slightly by a somewhat smaller object of art in the receptionist's office. On a small table near the door, sitting among brochures and issues of *Astrobiology*, is a small 10½-ounce can with an iconic red and white label, which, upon close inspection, reads "Campbell's Primordial Soup." It's a science joke, and a good one, although its referent—the idea that life on Earth began in a warm brew of amino acids—may have passed its expiration date. One flight up and just down the hall, two scientists have other ideas about the conditions that led to life, and they are working to simulate those conditions—as it happens, in a chamber about the size of a soup can.

ALLAMANDOLA

Lou Allamandola is a senior scientist in the Space Science and Astrobiology Division, and the founder and director of NASA Ames's Astrophysics & Astrochemistry Laboratory. He is tall and bespectacled, and were you to see him outside his lab, you might guess him to be a college basketball coach. His office is sunlit, its several tables and bookcases agreeably cluttered with drafts of papers to be proofed, a well-thumbed copy of *The CRC Handbook of Chemistry and Physics*, an orrery (that mechanical model of the Solar System, like a Calder sculpture, with marble-sized planets moved by gears), and many pictures of family. On one bookcase, curiously, are corked champagne bottles, with scribbling on their labels. Dr. Allamandola explains that when someone on his research team makes a discovery, the team celebrates, and the writing on the label is the name of whatever it was they discovered. Listening to Allamandola speak, an untutored ear might guess his origin to be Brooklyn or Queens—a fair try, as it happens. He was raised in the Italian section of Greenwich Village, and his inflections have survived unscathed through many years in California and a long stint at Leiden University in the Netherlands.

As Allamandola explains, a great deal has changed in our understanding of what is called interstellar chemistry—

beginning with the realization that there is such a thing. As recently as twenty years ago, most astronomers thought that interstellar space, except for an atom of hydrogen here and an atom of helium there, was barren. Some scientists conjectured about the existence of interstellar ice, but few took those conjectures seriously. Astronomers could see interstellar dust, but no one knew what it was made of, and most were quite sure it could not be large molecules. In those vast spaces, there simply would be no way for atoms or small molecules to find each other, and even if they did, they'd be torn apart immediately by ultraviolet radiation from nearby stars.

Now, though, we have a better understanding of what's in space and how it got there. The story goes something like this. The elements necessary to familiar life—carbon, nitrogen, and oxygen—are forged deep inside certain stars, and late in the star's life they are thrown off the star's surface. By that time they've been sufficiently mixed to form simple molecules like acetylene and carbon monoxide, as well as dust particles of carbon and silicates. All these are pushed into space—more specifically, the near vacuum that astronomers call the interstellar medium, an extremely diffuse gas of hydrogen and helium. There the molecules and dust particles are bathed in ultraviolet radiation, bombarded with gamma rays and subjected to more chemical reactions. Some of the molecules accrete on the surfaces of the dust particles.

The ultraviolet radiation destroys the smaller molecules, but others, along with the larger molecules and dust, are pushed into molecular clouds—these being vast regions of relatively dense gas and dust some light years across. Within the clouds, cold refractory grains block out enough ultraviolet radiation that simple molecules can survive and continue to form. It is cold inside a molecular cloud—a chilly 10–50 degrees above absolute zero— and the molecules condense onto the surface of the grains as ices. Confined to the very small surfaces and crowded against other molecules, they have another chance to engage in chemistry, and many do.

Meanwhile, on a much larger scale, part of the cloud itself becomes unstable, and areas with greater concentrations of matter begin to condense and collapse. Once a concentration has broken free from the other parts of the cloud, it is called a protostar. As more matter is drawn inward and releases kinetic energy, the star begins to burn. Some accreting matter is pulled into a slowly turning disk around the star, where it slowly coalesces into planetesimals, and then planets, moons, asteroids, and comets. Mixed with that matter are the grains coated with ices. They adhere to others like themselves, in time growing into ever-larger mixes of ice and organic material, and become comets. Early in a solar system's history they are plentiful, and many strike planets and moons, delivering great quantities of water and organic material, and setting the stage for life.

Many details of this picture are new. As recently as the 1970s, Allamandola says—despite the common experience of breath condensing on cold glass and the common knowledge that raindrops get their start by forming around particles of dust or salt— no one thought ices would condense on grains. When asked the reason for the doubt, he pauses. For a moment he seems about to speak of long experience with the intractability of the scientific establishment. But then he just shrugs and says, "People are people."[1]

A RECIPE FOR A COMET

Astronomers know what interstellar clouds are made of because light passes through them; and by using interferometry and analyzing the spectrum of that light, they have detected many types of molecules. Comets, though, present a more difficult problem. Many of us have heard astronomers call large comets "dirty snowballs," fragments of ices and rock a few miles across, covered with a dusting of organic matter. As it happens, that's only a rough description, as no one is sure exactly what kind of ice and exactly what kind of organic matter. Allamandola says, "We don't know yet, really, what comets are made of."[2]

Finding out won't be easy. Being mostly solid bodies, comets resist interferometry. In 2003, it occurred to Allamandola and his colleague Doug Hudgins that there was another way. If they could make a comet from scratch, they could study *it*. They took a sample chamber, removed most of its atmosphere, and froze what was left to a temperature near absolute zero, thus creating a fair representation of deep space. They introduced into the chamber a few simple molecules that might be found in a star's outflow, and turned on a lamp (representing nearby stars) that bathed the molecules in ultraviolet radiation. Then they waited. They were not expecting much, and they were certainly not expecting what happened. The molecules combined, split, and recombined, and before long the chamber contained some very complex molecules, many of which were prebiotic.

Astronomers using interferometry have learned that among the large molecules spread through the interstellar medium are polycyclic aromatic hydrocarbons (or PAHs). Under an electron microscope, PAHs look like pieces of chicken wire; and

being resistant to ultraviolet radiation and other unpleasantness, they are at least as hard to pull apart. Since they seem to
be almost everywhere in space, there was every reason to expect
they would find their way into comets. When Allamandola and
Hudgins included them in the mix, even more types of molecules
were produced. They were much more complex than those made
without PAHs, and again, many were prebiotic.

Allamandola wondered how much chemistry might go on
inside a comet as it traveled along its orbit around the Sun. It
was possible, he thought, that as surface layers of ice and organics boiled off, producing the comet's tail, ices inside might melt,
refreezing as the comet departed the Sun's vicinity, melting again
on its return, and so on, thereby enabling still more complex
chemistry. By way of testing the hypothesis, Allamandola and
biochemist David Deamer (of UC Santa Cruz) took ices formed in
the chamber and dropped them into liquid water, and then put a
drop of that water under a microscope slide. To their amazement,
they found circular structures that looked like red blood cells. Of
course, they were not cells, and Allamandola and Deamer were
careful to call them "vesicles," but they bore more than a passing
resemblance to cell membranes, with several lipid layers separating inside from outside. There was another surprise. Under
ultraviolet light, the light that they used to mimic conditions in
space, the vesicles *fluoresced*.

The vesicles themselves were significant, of course, because
they suggested the possibility that such structures were precursors to cells, and—more provocatively still—that those structures originated not on Earth, but in interstellar space. The
fluorescence was significant because it meant that the vesicles
might absorb more energy in the ultraviolet range of the spec-

trum than they release in the visible range. If they do, they would enjoy an energy surplus that, like the bacteria taking advantage of the sulfate-producing reaction in the vicinity of hydrothermal vents, they might put to other uses, like synthesizing molecules. That fluorescence would have been doubly useful early in Earth's history, before an ozone layer shielded its surface from ultraviolet radiation, when it might have performed a neat biochemical jujitsu, redirecting some of that energy toward its own uses and at the same time rendering it harmless.

Most of the compounds that Allamandola and his team have made in their chamber have been found in meteorites. In fact, scientists had thought that the aromatic hydrocarbons in meteorites were created when simpler hydrocarbons were heated to high temperatures by entry into the atmosphere. The evidence now is that they, with much else, were made elsewhere. And the upshot is a real Kuhnian paradigm shift.

Many scientists had long assumed that the chemistry that enabled life on Earth began here, as it were, "from scratch," and many had assumed that prebiotic molecules could be formed only on a warm planet with an atmosphere and water. But what Allamandola and Hudgins found in their sample chamber made them (and many others) reconsider those assumptions. It was entirely possible that when comets brought water to Earth, they also brought the chemical compounds necessary for life. "Perhaps Darwin's 'warm little pond,'" wrote Allamandola and Hudgins, "is a warmed comet."[3] But *only* perhaps. Allamandola is careful to say that they have a long way to go to show that the ingredients of the primordial soup, let alone life, may have been delivered here via comets. He is also careful to say that there is no evidence that life itself originated in molecular clouds.[4]

Another scientist, though, has argued, if not that life began in such clouds, then that at some moment in an unimaginably distant future, they may be the place where it makes its last stand.

LIFE IN A DISTANT FUTURE

In 1923, British geneticist J. B. S. Haldane published a paper called *Daedalus: or, Science and the Future*; and in 1929, John Desmond Bernal published a monograph called *The World, the Flesh and the Devil: An Enquiry into the Future of the Three Enemies of the Rational Soul*. At about the same time, Jesuit priest and philosopher Pierre Teilhard de Chardin was developing his own account of the long unfolding of the material cosmos, an unscientific (albeit quite poetic) description of the long ascendancy of life. All these works, appearing in the first half of the twentieth century, were predictions of the future of humanity and life. In 1977, physicist Jamal Islam published an article that predicted the future of the physical (and nonbiological) universe. It was left for a physicist with a philosophical bent to pull these strands together.

In 1979, Freeman Dyson, a mathematician and theoretical physicist working at the Institute for Advanced Study in Princeton, New Jersey, published a brief work called "Time without End: Physics and Biology in an Open Universe." In it, he proposed that the basis of consciousness may be structure, and that structure may require no particular form of matter. If this is true, then it follows, he claimed, that "life is free to evolve into whatever material embodiment best suits its purposes."[5] He went on to propose a means by which life might survive in the future universe that current cosmology predicts, not for a mere billion years (when Earth's surface will be too hot for familiar life) or even a trillion years (when most Sun-like stars will have burned themselves out), but for a literal eternity.

Fifty years earlier, most scientists would have considered such a prospect flatly impossible. In the late nineteenth century, astronomers reasoned that since all closed systems eventually reach thermodynamic equilibrium, and since the universe is a closed system, there would come a time when every part of space settled into the same very low temperature—a condition astronomer Arthur Eddington called the "heat death" of the universe. Because life depends on warmer and colder places and the transfer of heat between them, it would mean the death of all life as well. But the premises of the heat death were soon called into question. In the late 1920s, astronomers confirmed that galaxies in all directions were moving away from our own, and that the universe—and space itself—was expanding. As long as the universe continued to expand, it would never reach thermal equilibrium.

In 1979, Dyson held with evidence that the "observable universe"—that is, the part of the universe we can observe, an imaginary sphere 84 billion light years across and centered on us—was expanding at a faster rate than was the entire universe. Its relatively faster expansion meant that although all galaxies were receding from us, just inside the edge of the observable universe more and more of them were coming into view. Although the universe that Dyson imagined was growing emptier for a given volume of space, its *total* volume was increasing, and that was good news for life and intelligence. "No matter how far we go into the future," he wrote, "there will always be new things happening, new information coming in, new worlds to explore, a constantly expanding domain of life, consciousness and memory."[6]

It was this "open" universe that would supply habitats to the sort of weird life that Dyson imagined. Although the heat death was no longer a threat, an ever-diminishing supply of usable energy in an ever-colder universe would require organisms to practice frugality, and Dyson suggested that they might do so by slowing their metabolisms. There would be challenges. Because thought is a product of metabolic processes, intelligent organisms would be obliged to slow their rates of consciousness. More generally, all organisms, even very cold ones, would overheat if they did not generate waste heat away from themselves. And because the efficiency of any such radiator drops off much faster than does its metabolic rate, overheating, ironically enough, would become a real danger. But Dyson conjectured that overheating might be avoided if the organisms lowered their metabolic rate as averaged over time, and that they might do so by spending a large part of their lives in hibernation.

Although Dyson admitted that he could not imagine such organisms in detail (he could not, for instance, know whether there were functional equivalents of muscles or nerves), he noted that most biologists would be hard-pressed to imagine a cell if they had never seen one. He could, though, conjecture as to their general nature. He imagined a very distant future when most matter would have fallen into black holes, and yet he conceived a way for life to continue. "If it should happen, for example, that matter is ultimately stable against collapse into black holes when it is subdivided into dust grains a few microns in diameter," Dyson wrote, "then the preferred embodiment for life in the remote future must be . . . a large assemblage of dust grains carrying positive and negative charges, organizing itself and communicating with itself by means of electromagnetic forces."[7] Life, then, would not be *in* a molecular cloud; rather, it would have *become* a molecular cloud.

STARS

The universe used to be a comfortable place. The idea that it is *un*comfortable—that most nonterrestrial locales are barren and inhospitable—gained widespread acceptance only recently, in the early twentieth century. Before then and since ancient times, many astronomers, cosmologists, and philosophers assumed that the universe offered pleasant habitats in abundance. One rationale for a well-populated universe is a long philosophical tradition claiming that celestial bodies would be "wasted," were it not for observers. Johannes Kepler, for instance, reasoned that since Jupiter's moons cannot be seen from Earth with the naked eye, they must be meant for Jovians. In 1693, English theologian Richard Bentley enlarged the argument: "As the Earth was principally designed for the Being and Service and Contemplation of Men; why may not all other Planets be created for the like uses, each for their own inhabitants who have life and understanding."[8] Bentley's idea of an Earth made for us, even carrying the authority of Genesis, may seem faintly risible. But it's hard not to admire in his words an *absence* of self-centeredness as well—in the allowance for the possibility, if not probability, of other worlds and other beings.

Among possible habitats, some natural philosophers counted not only planets and their moons, but *stars*. Two thousand years ago, in his satirical work *True History*, Assyrian rhetorician and satirist Lucian of Samosata imagined that the Sun itself might be inhabited. As recently as the eighteenth century, no less distinguished a personage than astronomer Sir William Herschel, discoverer of the planet Uranus, supposed that the part of the Sun we see from Earth is something like our aurora borealis, and that beneath it was a layer of dense clouds that shielded denizens of the surface from our sight. Herschel's son John suspected that the ephemeral formations we call solar flares—and he called (rather more lyrically) solar willow leaves—might themselves be living creatures. No one gives much credence to these ideas now, but there have been more recent speculations of means by which life might survive in stars—or at least, certain kinds of life and certain kinds of stars.[9]

WHITE DWARFS AND BLACK HOLES

The current era in the universe's history is a relatively active period when stars are forming, living, and dying. It will end when the universe is 10^{14} years of age and all stars have cooled and faded to stellar remnants—white dwarfs, brown dwarfs, neutron stars, and black holes. This begins a very long and relatively quiet period lasting until the universe is 10^{28} years of age, its stillness interrupted only when two white dwarfs collide and create a supernova explosion that for a brief moment brightens an otherwise darkened galaxy. The period would seem utterly inhospitable to life, and perhaps it will be. Nonetheless, in 1999 astrophysicists Greg Adams and Fred Laughlin, taking a page from Dyson, imagined life in the atmospheres of white dwarf stars.

Some background is necessary. A star like the Sun generates energy by slowly fusing the light nuclei of hydrogen to the heavier nuclei of helium, and it does so as long as it has hydrogen left to fuse. During most of its lifetime, such a star is a balance of forces: the superheated gases of its interior pushing outward and the mutual gravitational forces of its parts pulling inward. When the star exhausts its hydrogen fuel and cools, the outward-pushing pressure diminishes, and the inward-pulling force overwhelms it, becoming so strong that the shell of electrons surrounding each atomic nucleus is squeezed, restricting each electron to a "cell" with a volume thousands of times smaller than the volume the electron would otherwise inhabit. The only thing that prevents the star from further collapse is outward pressure of the electrons against the walls of their cells—the electron degeneracy pressure. The star itself has become a white dwarf, a sphere about the size of Earth.

White dwarf interiors are unimaginably dense—10^{14} grams per cubic centimeter—but their atmospheres, so Adams and Laughlin think, would allow mobility. Those atmospheres contain oxygen and carbon, and although they are quite cold, they are warm enough that they would allow those chemicals to interact in interesting ways. White dwarf atmospheres gain what heat they have by collisions with particles of dark matter, a process that will continue until the dark matter is exhausted, when the universe is 10^{25} years of age. Over such a span—100 billion times as long as it took for life to appear on Earth—even slow-moving molecules will have time to join in every conceivable pattern, including those necessary for life. The longevity of a stable environment, Adams and Laughlin argue, implies that life in white dwarf atmospheres is more than possible; it is likely. They also note that it would necessarily be a life quite unlike our own. In accordance with Dyson's notions of energy conservation, metabolisms and rates of consciousness would be very slow. An intelligent creature living in a white dwarf atmosphere might take a thousand years to complete a single thought.

Still, such beings would not be immortal. When the universe is 10^{40} years old, even protons will have evaporated into a diffuse radiation, and the only remaining stellar remnants will be black holes. Their radiation will provide all the warmth available anywhere, and it will be precious little. A black hole with a mass a few times that of the Sun would have a temperature of one ten-millionth of a degree above absolute zero. But again, given enough time, matter and energy will arrange themselves into a great many forms, some of them likely to be living. Adams and Laughlin suggest that because the era of black holes is extensive, lasting until the universe is 10^{100} years old, life near a black hole event horizon (that theoretical boundary around a black hole beyond which no light or other radiation can escape) would have a long time to take hold and a very, very long time to evolve into complex forms—albeit, owing to a deficit in available energy, not intelligent ones.

This sort of speculation at present is untestable. But even the more modest hypotheses in the previous chapters present difficulties. As should be clear by now, the search for life else-where meets with problems because it requires life detection experiments. Whether performed with ground-based or space-based telescopes, unmanned spacecraft in orbit around planets or moons, or rovers and submersible robots, these experiments must be based on designs. Designs in turn must take into account aspects of microbiology, evolution, and planetary science. And because (as we've seen) scientists' knowledge of all these fields is incomplete, they can't be sure what makes for a good design, and they can't be sure what to test for to begin with. The same lack of knowledge hampers and vexes even theorizing about life.

In the late twentieth century, a number of scientists realized that the problems involved in designing instruments and agree-ing on what to test for could be leapfrogged, and the search for extraterrestrial life could continue apace—if only all concerned were prepared to make two assumptions: one, that some subset of extraterrestrial life was at least as intelligent as humans; two, that it had the means and the will to communicate across inter-stellar distances. As it happens, there is a group of scientists who have learned to live and work with both assumptions. These are, of course, the radio astronomers and other researchers involved in the search for extraterrestrial intelligence, or SETI.

CHAPTER SEVEN

Intelligent Weird Life

S ETI has been with us since 1959, the year that physicists Giuseppe Cocconi and Philip Morrison published a paper in the journal *Nature* outlining an in-depth strategy by which radio telescopes might be used to detect the communications of extraterrestrials.[1] Even for *Nature*, a journal known to publish work that approached the scientific fringe, it was audacious, beginning with what, it must be said, was a rather startling supposition:

> We shall assume that long ago [extraterrestrial civilizations] established a channel of communication that would one day become known to us, and that they look forward patiently to the answering signals from the sun which would make known to them that a new society has entered the community of intelligence.

The paper's conclusion, no less startling, was a call to action: "The presence of interstellar signals is entirely consistent with all we know now, and . . . if signals are present the means of detecting them is now at hand."

Cocconi and Morrison did not know that at that same time, the call had already been answered—or at least heard. A young radio astronomer named Frank Drake, then working at the National Radio Astronomy Observatory at Green Bank, West Virginia, suggested that it was possible to use the facility's 26-meter receiver to detect artificial radio signals—that is, signals sent deliberately by someone. Drake made a case to the observatory's director. First, such a project would cost next to nothing. He needed only a narrowband receiver and a parametric amplifier, and he could build both for $2,000. Further, the equipment could do double duty because the narrowband receiver could also search for the splitting of spectral lines in a magnetic field—a phenomenon known as the Zeeman effect. To Drake's great surprise and pleasure, the director agreed. Drake acted quickly, and chose as targets two Sun-like stars: Epsilon Eridani and Tau Ceti. At the moment he turned the system on, he received a very strong signal. It was cause for a few hours of cautious excitement, but a false alarm. Otherwise, the only sound that came from the loudspeaker at Green Bank was static. Nonetheless, Drake had spurred the interest of other astronomers.

In 1961 an informal meeting was held at Green Bank, its purpose to address questions associated with interstellar communication. There were many such questions, and Drake realized that they could be arranged hierarchically within an equation that, conveniently enough, might provide the meeting with an agenda. Now called the "Drake Equation," it is a set of seven unknowns ranging from the physical (the rate of star formation) to the social (the longevity in years of a technological civilization).* Replacing the unknowns with numbers yields an estimate of the number of civilizations in the Milky Way galaxy presently capable of interstellar communication.

By the time the conference ended, its members had estimated that the number of civilizations in the galaxy ranged from fewer than 1,000 to 1 billion, with Drake himself putting it at 10,000. The reason for the rather impressive range is that one factor was the longevity of such civilizations, and how long civilizations might last seemed sheer guesswork. Of course, the discussion—and the guesswork—has continued since. Generally, those advocating SETI projects have made the case for the existence of many extraterrestrial civilizations, and those dismissive of SETI have argued that there are probably few, if any.

The position of the second group, roughly speaking, is this. Complex, larger cells with nuclei and various internal structures—the kinds of cells that make possible plants and animals—did not appear until 2 billion years after the first life— archaea and bacteria—had established itself. The relatively late arrival of complex life suggests that it does not and perhaps cannot develop easily. In all the 30 billion or so species that have crawled, swum, and fluttered, none have developed intelligence on par with *Homo sapiens sapiens*—a fact suggesting that the

---

* The Drake Equation: $N = R^* \times fp \times ne \times fl \times fi \times fc \times L$. R^* is the number of stars that might host planets formed annually in the galaxy; fp is the fraction of those stars that actually have planets; ne is the fraction of those planets that are Earthlike—that is, having an environment suitable for life; fl is the fraction of Earthlike planets on which life actually appears; fi is the fraction of Earthlike, life-bearing planets on which intelligence appears; fc is the fraction of intelligent beings that develop the desire and ability to communicate with other worlds. L is the longevity of each society in the communicative state.

survival value of intelligence is no more or less than the survival value of features like, say, feathers or an exoskeleton or a prehensile trunk. It is true that some animals do manage a rudimentary technology, if technology is defined as toolmaking. But even if a species develops more advanced technology, unless that development is guided by scientific method it will progress slowly and fitfully, and it may never produce radio telescopes. Moreover, this line of thought continues, among all the civilizations in human history, scientific method arose only once, in western Europe in the late sixteenth and early seventeenth centuries. It follows that the chance for its appearance in any civilization (whether terrestrial or extraterrestrial) is slim. For all these reasons, those dismissive of SETI believe that although the universe may be full of life, almost all of it is likely to be microscopic, the equivalent more or less of archaea and bacteria, and looking for extraterrestrial civilizations is a waste of time.

The position of SETI advocates, again roughly speaking, is this: Since the first life on Earth appeared only a few hundred million years after the planet's crust hardened—that is, about as soon as life was possible—it was likely to appear anywhere as soon as it was possible. Intelligence has obvious survival value, and it has appeared on Earth in humans and to some degree in chimpanzees, dolphins, and several other animals. It may appear on many other life-bearing worlds as well. In time, intelligence is likely to develop science, technology, and radio telescopes. In the final analysis, say SETI advocates, we don't know whether extraterrestrial civilizations exist, and the only way we might discover them is with SETI programs. SETI, its advocates maintain, is a small investment for what could be a very big payoff. What exactly do they think that payoff may be? At the very least, the knowledge that others share the universe with us. And at most, the possibility that they possess the wisdom and experience to show us how to meet and overcome the many challenges our civilization now faces.

SETI STRATEGIES

Since 1961, there have been more than sixty separate efforts.[2] Most of these have followed Drake's lead and Cocconi and Morrison's recommendation that the easiest, cheapest, and most-likely-to-succeed search would be one that listens for radio signals. A few efforts—most notably the one led by Paul Horowitz at Harvard—have fitted an optical telescope with light detectors called photomultiplier tubes that can register a light pulse sent by a powerful laser. The justification for both strategies is straightforward. Both radio waves and lasers travel at the speed of light (that is, as fast as anything can travel), they can carry a great deal of information, they can be readily identified against the universe's natural background radiation, and they are vastly cheaper than alternatives like robotic probes. These are advantages that any technically minded civilization would notice, and any technically minded civilization wishing to communicate across interstellar distances would put to use—or so SETI thinking goes.

In fifty-odd years, there have been more than a few honest mistakes, one hoax, and one possible signal, the last never repeated but still unexplained.* The searches have made broad

---

* On August 15, 1977, Jerry R. Ehman detected a strong narrowband radio signal while working on a SETI project associated with the Ohio State University. The signal matched the expected signature of an interstellar signal, and it lasted for a full 72 seconds. Ehman circled the signal on the computer printout and wrote "Wow!" in the margin. It was not detected again, and so has not sat-

but reasonable assumptions about the signals' senders. Some searches were targeted at stars, with the implicit supposition that the senders would live in a solar system. Others, called "sky surveys" or "all-sky surveys," allowed that the senders may have set up shop in the vast reaches of interstellar space.

Natural sources of radio waves (like our galaxy and the Earth's atmosphere) broadcast over a broad range of frequencies. SETI searches assume that a civilization wishing to communicate via radio would use a signal clearly distinguishable from such sources. Accordingly, most searches have tried to detect narrowband signals lying in a range of frequencies that is mostly quiet—a sliver of the microwave region of the electromagnetic spectrum between the emission line at the 21-centimeter hydrogen wavelength and, just a tweak of the dial away, the 18-centimeter emissions of OH, the hydroxyl radical. There may be another reason for the choice, not strictly scientific. It so happens that hydrogen and hydroxyl can be combined to make water. People who listen for messages from the stars can be a fairly romantic lot, and Barnard Oliver, a leading figure in SETI since its beginnings, called the sliver "an uncannily poetic place for water-based life to seek its kind."[3] He and others after him began to call it the "water hole." If the medium isn't quite the message, so their thinking goes, it can at least tell us where to look for the message.

Most searches have assumed that the first signal detected will not be the transmission itself but the "carrier" that underpins that transmission. It would be a simple sustained tone or a series of short, intense pulses conveying no information beyond its self-evident artificial nature. The transmission—the actual

isfied the requirement of the scientific method that data from a given experiment be reproducible.

message—would be far weaker, and to receive it we would need to develop more sensitive receivers.

In the half century since the meeting at Green Bank, much has happened to SETI. Funding has disappeared and reappeared, and serious doubts have been raised as to the existence of extra-terrestrial civilizations. Yet two factors of the Drake Equation— the rate of formation of stars that might host planets, and the fraction of stars that do host planets (both unknowns in 1961)— have come within the scope of observation. Astronomers now estimate that between ten and twenty stars that might host plan-ets are born every year, and astrophysicist Geoff Marcy, whose research group has discovered the majority of extrasolar plan-ets, believes that between half and three-quarters of all stars are accompanied by a retinue of planets. NASA's Kepler space observatory mission team is conducting a survey that will fully address the second term and partly address the third term of the Drake Equation—the fraction of stars with planets and the frac-tion of those planets that are Earthlike, respectively. And now SETI itself is old enough to have a history, and its own historical places. At Green Bank, in the meeting room where Drake first scribbled his equation, there is a commemorative plaque.

Might any of the assumptions made by SETI cause a search to overlook intelligent weird life? Probably not. Although the chemistry and biochemistry of intelligent weird life might dif-fer, its physics would be our physics.* A technical civilization of weird life would recognize the advantages of communicating via radio waves or light pulses as easily as we do, and although its radio transmitters and lasers might be manufactured with a met-allurgy derived from a rather exotic chemistry, they would work

---

* We will get to ideas of even weirder life—with different physics—in Chapter 9.

as well as ours. Certain sorts of intelligent weird life—those that, say, grew thirsty for liquid methane—might not regard Oliver's water hole as particularly poetic (assuming of course, they knew what poetry was), but they would nonetheless recognize it as the best range of wavelengths for sending and receiving signals.

THE CHANCES FOR INTELLIGENT WEIRD LIFE

Another question arises—rather nearer our interests here: What are the chances that intelligent extraterrestrials would have a biology radically different from our own? We can guess. Certainly it is easy enough to dust off the Drake Equation again and rejigger it for weird life. The third unknown in the equation—the fraction of planets that are Earthlike (that is, having an environment suitable for life as we know it)—might be altered (narrowly) to represent the fraction of planets that are, say, Triton-like if we're looking for nitrogen drinkers or Titan-like if we're looking for methane drinkers. As we saw in Chapter 5, planetary scientist Jonathan Lunine makes this point regarding Titan in particular: "A positive answer would force the third term in the Drake Equation—ne—traditionally defined in terms of the environment of our own Earth, to be radically expanded."[4] As it happens, an expanded definition of that term yields something else: an answer to what SETI calls the "where are they?" question.

The question, which arose some twenty years after SETI began, challenged one of its central assumptions. Distances between stars are great, faster-than-light travel is impossible in the universe Einstein gave us, and travel at speeds approaching the speed of light would be prohibitively expensive in terms of energy. For these reasons, most SETI practitioners had assumed it unlikely that any civilization, no matter how advanced, would practice routine interstellar travel, and this assumption justified the emphasis on communication with radio telescopes and lasers. Physicist and Nobel laureate Edward Purcell was speaking for many when he remarked, "All this stuff about travelling around the universe in space suits . . . belongs back where it came from, on the cereal box."[5] But it seemed that the ingenuity given to breakfast product packaging was, in the view of some anyway, lacking as regards ideas for the possible varieties of interstellar travel.

Although the distances between the stars are vast, it is also true that the galaxy is old, and there are Sun-like stars 5 billion years older than the Sun. There is the possibility of Earth-like planets of the same age. What this means is that there has already been ample time for a star-faring civilization—even a slowly moving one—to colonize the galaxy many times over. In the mid-1970s, scientists Michael Hart and David Viewing took another look at the energy requirements and concluded that slower interstellar travel might be managed with nuclear propulsion at a cost that, for the civilizations they were imagining, would be reasonable.[6]

Hart calculated that civilizations with the ability to travel at a relatively modest velocity of one-tenth the speed of light could have spread through the entire galaxy in only 1 million years. He acknowledged any number of reasons such civilizations might not do so. Perhaps some preferred a more contemplative lifestyle; perhaps some simply lacked the interest; perhaps some destroyed themselves before they had a chance. But SETI was postulating thousands of civilizations, and hundreds of thousands over time. If only one of them had spread out across the Milky Way, the evidence would be unmistakable—if not a family of squamous purple ovoids moving into the apartment downstairs, then at least detectable radiation from a spacecraft's drive, or an artifact on Earth or the Moon. SETI researchers, discounting the claims of authors like Whitley Strieber and Erich von Däniken as unable to withstand rigorous proofs, agree that there is no such evidence. That we see no evidence, so argued Hart, means there is none, and that we are probably here alone.

The thinking was reasonable, and for SETI practitioners, the conclusion sobering. The "where are they?" question pushed SETI to what one practitioner called "a major crisis of identity and purpose."[7] The crisis led in turn to an intellectual houseclean-ing, forcing the formulation of new ideas and the clarification of old ones.[8] The strongest argument against Hart and Viewing's conclusion (and the best answer to the question) was one of the simplest. Phillip Morrison suggested that extraterrestrials had not colonized the galaxy simply because "something limits every growth."[9] In other words, we could not know exactly what pre-vented the colonization, but we could be sure that something did, and we could be sure that something always would.

But there were other arguments too. A civilization might have colonized the galaxy and still be invisible to us, if members of that civilization were a certain kind of weird. Recall Lunine's conjecture that there may be many, many more Titans than Earths, and consequently much more Titan-like life than Earth-like life in the galaxy. If we jump over one or two other unknowns in the Drake Equation, we might conclude that there is a greater possibility for a civilization to arise on a Titan-like planet (or moon) orbiting a red dwarf star than on an Earthlike planet orbiting a Sun-like star. We might further suppose that intelli-gent beings that had evolved on such a world and were interested in interstellar colonization would, like many a colonist, seek the familiar. So one answer to "where are they?" is that they are (or were) in red dwarf star systems.

The Drake Equation could also be altered (widely) to represent that rather larger fraction of planets and moons and comets that would support any and all weird life. We could argue in general terms that if the arrangement of matter in our Solar System is typical—that is, with most of the real estate farther out from its sun and well outside the traditional habitable zone—then "cold" weird life has more places to get a foothold, in fact about a million times more. If the other factors in the equation remain unchanged, there is a correspondingly greater chance that the signal will be from a weird intelligence. Of course, all the other factors probably do not remain unchanged.

SHOSTAK

The most ambitious SETI project at present is privately funded, and conducted jointly by UC Berkeley and the SETI Institute in Mountain View, California. The senior astronomer at the SETI Institute is Seth Shostak. He has been involved with SETI for more than twenty years, and since 2003 he has served as chair of the SETI Permanent Committee aligned with the International Academy of Astronautics. Shostak has probably given more time to thinking seriously about extraterrestrials than has anyone else alive. He is the one the news outlets call when someone reports a signal, and if and when a signal is detected, he will in all likelihood be the one you'll see on the talk shows explaining narrowband frequencies and drift scans.

Shostak, now in his sixties, has the relaxed and cheerful

manner of a man who long ago made peace with any doubts about his career choice. He can answer skeptics in sound bites and statistic-laden twenty-minute lectures, and he can do either with an ease that suggests he knows something they don't. In fact, he probably does. It has to do with the chances of detecting a signal.

Until the UC Berkeley–SETI Institute collaboration, the most ambitious SETI effort was Project Phoenix. It lasted nine years and listened for signals from stars inside an imaginary bubble centered on Earth and extending outward 150 light years—750 stars in all. But there are 200–400 *billion* stars in the Milky Way galaxy. Consider Frank Drake's estimate that distributed randomly among those stars there are 10,000 communicating civilizations. If we sifted through the static at Project Phoenix's rate, we would expect to spend about 98,000 years before finding a signal. But—and this is a point Shostak makes often—the speed at which signals can be processed increases along with computing power, and in accordance with Moore's law,* computing power doubles every eighteen months. Assuming the trend will continue (and there's no reason to think it won't), then in the next twenty years the Allen Telescope Array, with regular upgrades, can be expected to have examined 1 million stars. If Drake's estimate is right, then odds are good that by 2030, SETI will have detected at least one extraterrestrial civilization.

---

* In 1965, Intel cofounder Gordon E. Moore described a trend whereby the number of transistors that can be placed inexpensively on an integrated circuit doubles approximately every two years. The figure of eighteen months used by Shostak (and many others) is derived from Intel executive David House's estimate of the period in which chip performance doubles. (Kanellos, "Moore's Law")

THE NATURE OF EXTRATERRESTRIAL CIVILIZATIONS

What might that civilization be like? Shostak suspects that although we cannot predict the "macroscopic structure and general demeanor" of the extraterrestrials who send a signal, we can make an educated guess as to their nature.[10] First, he thinks, the senders of any signal will have a technology that is thousands, perhaps millions, of years ahead of ours. To understand his reasoning, we need to back up a bit.

Cosmologists put the age of the universe at 13.7 billion years. Shrink that span to a single calendar year, and the galaxies form in May, the Sun and its planets form in mid-September, one-celled organisms appear on Earth in early October, and a month later the atmosphere begins to oxidize. The first dinosaurs hatch on December 24, and for several days they reign as Earth's largest land-dwelling organisms. On December 30, the first small warm-blooded animals scurry through the underbrush. Early on New Year's Eve, *Homo erectus* appears and disappears, and Neanderthal comes into view and vanishes. The rise and fall of civilizations; the vast migrations of peoples; the discoveries of continents; the struggles of nations and empires; the creation of art, philosophy, and religion; the development of science and technology; all that has been accomplished by human civilizations—everything that we call recorded history—appears in the final ten seconds of the year, roughly the time it took you to read this sentence. Of course, the clock doesn't stop here, and assuming that human civilization survives for another ten seconds on the calendar, we might reasonably suppose that its technology will advance as far beyond ours as ours is advanced beyond the reed boats of Mesopotamia.

For the sake of argument, let's define the current stage in

human history—from radio in the 1930s to the present glob-ally networked computer technology and communication, unmanned spaceflight to the nearer planets, and so on—as last-ing for the last one-tenth of a second on the cosmic calendar year. Since we have only one example to work with, we can't know how long it takes for life to arise, intelligence to develop, and civiliza-tions to emerge. But let's be conservative and suppose that no civilization anywhere in the galaxy could have arisen earlier than 200 million years ago, the moment on Earth when mammals first appeared. In other words, let's suppose that all civilizations in the galaxy arose *after* that moment. On the calendar, that's Decem-ber 26—six days ago. The odds of any civilization beginning at a random moment in the previous six days and reaching a stage of development equivalent to the current stage in human history at precisely the same one-tenth of a second are (the word is ines-capable) astronomical. Thus does the calendar in a single stroke dismiss the densely populated universes of much science fiction. The notion of scores of interstellar civilizations spread halfway across the galaxy—all with roughly similar technologies, warring with each other on bad days, and on better days negotiating trea-ties and trading recipes—is a fantasy, and likely to remain one.

The truth is likely to be far stranger. And lonelier. There may be civilizations that winked on and off in a few seconds on the calendar, civilizations that reached their height when Earth's most complex organisms were trilobites. They were too early for us, and we are too late for them. There is very little chance that a signal will come from a civilization that just lit up in the same tenth of a second we did. Any civilizations that might exist at this moment, and so any we are likely to hear from, have been lit for a while, and stayed lit. They will be much older than us and, we may assume, much more advanced technologically.

What can we say about such beings, and such civilizations? Shostak, for one, suspects they will be machines. His reasoning is based on predictions of our own future, and the assumption that most technological civilizations created by biological beings will reach a point in their history at which they will cede their power to artificial beings. We might add here that there is every reason to expect that beings with a weird biology—say, beings with a biochemistry based on silicon and ammonia (as long as they had an aptitude for technology)—would do the same.

In 1994, artificial intelligence pioneer Marvin Minsky penned a piece entitled "Will Robots Inherit the Earth?"—a rhetorical question, which the piece answered more or less in the affirmative. Experts in artificial intelligence expect that members of the human generation now in its twenties will live to see the day when computers match the computational power of the human brain. Since computer power increases exponentially, only fifty years after that day there will be computers or computer networks with the computational power of everyone on Earth. In any measurable way, they will be a lot smarter than we are, and we would be wise to admit it. If our past is any guide to our future, we probably will.

Many of us long ago ceded spelling and multiplication to the better judgment of software programs. If you checked my math a few paragraphs back, you probably did not use a pencil and paper. Stand on a busy street corner and look around at the pedestrians studying their iPhones and Blackberrys, and at the automobiles with drivers using GPS navigation, and you might well conclude that we've already become cyborgs, and we've managed to do it without subcutaneous computer chip implants. All of which is to say that insofar as there was a war with the machines, it was bloodless. The humans lost, and most of us don't seem to care.

As individuals, we relinquish our choices of restaurants (and best routes to get to those restaurants) to software programs. As a society, we relinquish our judgment on matters of greater consequence, like health care and finance, to programs and databases. There is no obvious reason we will not continue this practice; in fact, most likely we'll extend and enlarge it, sooner or later allowing computer networks to manage agriculture, trade, economic, and environmental policies—that is, everything on Earth that can be managed.

Both milestones cited in the preceding paragraphs (the development of a computer with the computational power of a human brain and the development of a network with the computational power of everyone living) are human centered, and there is no reason to expect computer development to stop or pause to acknowledge either. The rather more relevant marker, at least as described in 1993 by mathematician Vernor Vinge, is the moment when all networked computers exchange information at such a rate that they become self-aware. Then, Vinge said, they will begin to adopt motives and goals that the smartest among us simply won't understand.[11] The train will have left the station, and we'll be standing on the platform, wondering if they'll ever write home. Meanwhile, our electronic progeny will design their own progeny from scratch, and natural selection will give way to a Lamarckian, self-directed evolution that will be very, very fast, with generations succeeding generations in seconds and then milliseconds, its individuals fashioned with genetics, robotics, and nanotechnology for any and all purposes and any and all environments. And because they will be able to replace parts and download and upload memory as necessary, for most intents and purposes they will be immortal.

Suppose Vinge is right. Is it possible that with vast knowledge comes humility and a deep sense of history, and that our machine offspring will remember us—and perhaps even find a reason to respect us? More than forty years ago, science fiction author Arthur C. Clarke imagined extraterrestrials who, over millennia, traded worn-out bodies for machines and, many millennia later, worn-out machines for pure energy. Although they had become "godlike," Clarke writes, "they had not wholly forgotten their origin, in the warm slime of a vanished sea."[12] But we are, phylogenetically, a lot nearer one-celled organisms than are Clarke's godlike energy beings, and when we see warm slime washed up on the beach many of us feel no particular intimacy. It is as possible that they will be what H. G. Wells and many after him described as "something inhuman, unsympathetic, and overwhelmingly powerful."[13] Will they decide, as in so many science fiction movies, that we should be exterminated, or merely shuffle us off to some sort of global assisted-living facility and go on without us?

On all these questions we may take some comfort in the existence of the Singularity Institute, a think tank founded in 2000, whose members consider means to ensure that our machine offspring remain "human friendly." Its members work in and around Berkeley, which happens to be a few miles from the SETI Institute in Mountain View, which happens to be a ten-minute drive from NASA Ames. The future arrives in some places sooner than others, and it seems to arrive in northern California sooner than anywhere else.

WEIRD MACHINES

Let's return now to the question: Would senders of a radio signal be descended from beings with a weird biology? If they are machines, any answer might be moot. Following Shostak's reasoning, they would be designed from scratch by other machines that had been designed from scratch, and they would be designed for specific purposes. New designs would appear with each generation, and each generation might last a fraction of a second. Such beings would have no practical need to retain any vestigial physical characteristics, and the evolutionary break with their past might be clean. Even if an observer could examine such beings on the molecular scale, he would be unable to tell whether the biology or biochemistry of their ancestors was weird. Of course, the beings themselves would have great capacities for information storage, and would retain the knowledge somewhere. Some among them might have an interest in ancient history, and a few might harbor a sentimental need to recall their origins. But on the aggregate they would have no practical need for such memories, and it's possible they wouldn't care.

DECIPHERING A SIGNAL

Serious thinking about the "what if we receive a signal?" question has advanced further than you might suppose. There is a "SETI protocol" whose relevant paper is entitled "Declaration of Principles Concerning Activities following the Detection of Extraterrestrial Intelligence." Its signers have agreed, among other things "to inform the Secretary General of the United Nations as well as the public and the international scientific community, to the greatest extent feasible and practicable, of the nature, conduct, locations and results."[14] But what, exactly, would we be responding *to*? What might that message be?

Frank Drake has a recurring dream in which he detects a signal from a civilization some 20,000 light-years distant and is unable to decipher it. In his waking hours, he and his colleagues have considered the problems of understanding such a missive at length. Certainly its content would be constrained by the nature of our universe. Although radio waves travel at the speed of light, interstellar distances are very large. The first confirmed planet in a traditional habitable zone—Kepler-22b—is 600 light years from us, meaning that any message we receive tomorrow would have been sent about the time Henry V defeated the French at the Battle of Agincourt. The time lag might be inconvenient, but it is an inevitable consequence of the way our universe is, and civilizations wishing to communicate would have no choice but to adapt. They would design messages on matters that they judged important and useful, and they would transmit those messages with no hope of a quick response.

This was the expectation of Shklovskii and Sagan's *Intelligent Life in the Universe*, the primer of SETI, and remains the assumption of many practitioners. More specifically, Sagan suggested that there may be three overlapping signals: first, an unmistakable artificial beacon like a sequence of prime numbers; second, a primer to the language of interstellar discourse; and third, at the same wavelength but perhaps at a higher frequency, an "Encyclopedia Galactica"—that is, a catalogue of the civilizations known to the sender of the message, including descriptions of themselves and everything they know about the universe.

Marvin Minsky once posited a universal. Any intelligent species will use symbols simply because they are the only way to communicate a great deal of information economically. The word "apple" for instance, is made of five letters arranged in a specific sequence. It is far easier to reproduce those letters than an actual apple or a detailed image of an apple. Intelligent beings would recognize a kind of law of communication economics, and would abide by it. Any message sent by intelligent beings would be in a symbolic language. Moreover, it would be designed to be deciphered, and it would teach its recipients to decipher it with mathematics.

The reasoning here is twofold. First, anyone with the savvy to build a radio telescope must have learned enough mathematics to grasp the physics of electromagnetic radiation and enough three-dimensional geometry to design and build a high-gain antenna. Since the senders and receivers would both need radio telescopes, they would have this background in common, and both would know it.

Second, despite differences in their means of perceiving the world, even radically different minds are likely to share certain views. Physicist Max Tegmark notes that a cat seeing only black and white, a bird seeing four primary colors, a bee seeing polarized light, a bat using sonar, and a robotic vacuum cleaner—all would agree on whether a door is open or closed.[15] Radically different minds might also agree on more complex information, like that embodied by mathematics.

In fact, in the view of many (but certainly not all), mathematics is the very warp and woof of the cosmos, there waiting to be found, and waiting to be used as the basis for a universal language. The sum of 1 and 1, the number of degrees in a triangle, and the value of pi are the same everywhere in our universe. So are all mathematical theorems. This is why mathematicians speak of "discovering" a theorem, not inventing or creating one. Even supposing that mathematics can be used as a universal language, it does not follow that all of the hypothesized extraterrestrial civilizations use it. Certainly it is possible to imagine a civilization of poets and philosophers who manage with nothing beyond arithmetic. It is even easier to imagine one without engineering. But those are not civilizations that SETI expects to hear from.

COSMIC INTERCOURSE

It may surprise many that these ideas have progressed much further than theory, that linguists in our own galactic neighborhood have already designed languages built from mathematics. Dutch mathematician Hans Freudenthal, for instance, invented a simple language of long and short radio pulses for what his English translator called "cosmic intercourse."[16] Unless Freudenthal had a goal far more ambitious than that of most SETI programs, the unintentional double entendre is a lesson in humility for interpreters of Earthbound languages and, perhaps especially, for interpreters of signals from out there.

Inventor and software developer Brian McConnell has combined ideas from genetics and research in artificial intelligence to create another such language. In 2001, he argued that a message from extraterrestrials that could be deciphered might appear as a very long string of binary numbers—that is, ones and zeros—and they would present a visible pattern or structure, the equivalent of words and sentences. A decipherer would assume such a pattern was in there somewhere and begin her work by finding it. She would also assume that the message would begin by teaching mathematical symbols, and it would do so with examples. The equal sign, for instance, would be a certain sequence of ones and zeros, and a "not equal" sign would be another such sequence. Both would appear between other ones and zeros, but only the equal sign would appear between identical sequences.

Teaching other symbols and sets of symbols through more examples, McConnell's message would present a carefully staged curriculum, working on up through algebra, Boolean logic, and branching statements. Using bitmaps, it could teach with images in gray scale and color. By adding to the bitmaps a third dimension representing time, the message could use moving images. And with a set of moving images, it could convey an abstract concept. The idea of gravity, for instance, might be expressed by moving images of a falling weight, a planet in orbit around a star, two stars in mutual orbit, and a single star imploding. If the message attached the same sequence of ones and zeros to each of those images, it would identify them as representing the same phenomenon, and in all subsequent lessons it could dispense with the images and signify gravity more economically, with that sequence of ones and zeros.

If the message were something like an "Encyclopedia Galactica," McConnell thinks the main challenge to the message maker would be mediating its own knowledge with the lesser knowledge of the decipherer. The remedy would be a curriculum staged so carefully that when the decipherer reached a certain level in the message—a point at which that she could not decipher, despite all efforts—she would know that her capacity to understand had reached its limit.

You may see assumptions here. Perhaps the civilization that created the message does not conceive of spatial dimensions as we do. Perhaps that civilization does not conceive of time in such a way that it can be represented as a spatial dimension. Perhaps the SETI folk are too optimistic.

THE CHALLENGE OF UNDERSTANDING

Suppose, though, that we could communicate. How much might we understand? A similar question was addressed by philosopher

Thomas Nagel in his essay "What Is It Like to Be a Bat?" Nagel chose the sensory experience of bats from among many others because in his view, it is alien to our own. His answer to the titular question is that because we are not bats, we probably can't really know.[17] He may be right, but there is something defeatist in his answer, and it certainly runs counter to what may be a universal human impulse. We spend much of our lives trying to understand others and trying to make ourselves understood by them. On our best days, we are excited by communicating and understanding those who are truly different. In fact, much of art and literature attempts to communicate a particular awareness, sensibility, or emotional state, and some despairs of the possibility of such communication. This may be the very reason why Nagel's question is intriguing and challenging—too much so to leave to science fiction writers and spoilsport philosophers.

The large field of study called animal cognition is concerned with questions in areas of animal reasoning, problem solving, and language. Practitioners in the field study many animals, especially primates, cetaceans, and elephants. Some researchers are interested in consciousness, asking, in so many words, what it's like to *be* that animal. A great many nonscientists have asked the same questions; some have gone so far as to supply answers.

Numerous writers have tried to give expression to awareness that is nonhuman. There is a surprisingly large body of literature, for instance, that assumes the voice of dogs. Virginia Woolf penned an autobiography of the spaniel belonging to Elizabeth Barrett and Robert Browning; more recently there appeared a collection of poems written by poets using their dogs' voices (or at least what the poets imagined were their dogs' voices), with such refreshingly candid titles as "Are You Gonna Eat That?" and "Squirrel!"[18] Of course, as Nagel would remind us, not being dogs, we can't really know whether such renderings have any fidelity to the actual experience. All we can say is that they *seem* right. Dogs, of course, are near all of us phylogenetically, and many of us literally. It is possible that the gulf between human and canine experience is relatively narrow and easily bridged. A noteworthy attempt to bridge a wider gap—by someone who has given a great deal of thought to the creatures on the other side—is Edward O. Wilson's novel *Anthill*, parts of which are told from the point of view of an ant colony—a collective intelligence.[19]

Which brings us to John C. Lilly. In 1961, when Frank Drake and a staff officer on the Space Science Board of the National Academy of Sciences named J. Peter Pearman were planning the conference at Green Bank, they quickly came up with a fairly impressive guest list that included astronomers, chemists, and electronics experts. Drake joked, "All we need now is someone who has spoken to an extraterrestrial," at which point Pearman, taking the "terrestrial" in "extraterrestrial" to mean dry ground but otherwise not missing a beat, said, "John Lilly."[20]

Lilly was the director of the Communications Research Institute in the Virgin Islands, the communication of interest being that of bottlenose dolphins. In 1961, Lilly was fairly well known even outside academic circles, mostly thanks to a best-selling book called *Man and Dolphin*, which made the case that dolphins are as intelligent as humans, that they have a sophisticated language and perhaps an equally sophisticated culture. Lilly's work (and his capable promotion of it) exalted the dolphin in the popular imagination. It is thanks to Lilly that many of us regard the dolphin as wise, gentle, in intimate contact with nature, and, when called upon, heroic. Drake agreed with Pearman that Lilly might make an interesting contribution to the conference; he was among the invited, and his presentation, rich with anecdotes of dolphins demonstrating all manner of intelligence, was by all accounts enthralling. His stories of the challenges and exhilaration of communicating with dolphins also gave a hint of what might come with encountering a signal from the stars.

Some years later, Lilly designed a "communications laboratory," a half-flooded living room where humans and dolphins could communicate as equals and where they would develop a common language. That language failed to emerge, and Lilly came to believe the limitation was on his end. His solution was to attempt to make changes in his own mind. He began experimenting with LSD, ketamine, and a sensory deprivation tank, sometimes in combination, and "communed" with dolphins while using LSD. In the view of some, Lilly's work set back the study of dolphins by decades.[21] Carl Sagan mentioned Lilly's work in his best-selling book *The Cosmic Connection*, and he visited Lilly at the institute. But Sagan eventually concluded that Lilly's experiments lacked scientific rigor, and in time SETI and Lilly went separate ways.

There have been many other attempts to understand a very different awareness or consciousness. Some, interestingly enough, disregard scientific rigor altogether.

A TRAVELOGUE FOR PLANTS

Jonathon Keats is a San Francisco–based conceptual artist who would prefer to be called an "experimental philosopher." Many of his works explore the largely unmapped spaces separating law, philosophy, and science—places where few of us have thought to go. He used the legal framework of air rights to sell properties in other dimensions; he copyrighted his own mind, claiming that it is a sculpture that he created in the act of thinking; and—this of most interest to us—he filmed travelogues for plants. In March 2010 he created an installation of houseplants sitting on a floor beneath a video screen on which was projected a six-minute film loop of the sky over a part of Italy. By way of explanation, Keats deadpanned to *The New Yorker* that plants are more sensitive to certain parts of the light spectrum, and the makeup of the spectrum differs according to location.[22] Plants may never visit Italy, and they would be more interested in the sky than in sites favored by more ambulatory tourists. Thus, a good travelogue for plants—good from the plants' point of view, that is—would focus on the sky.

On the subject of his own artistic development, Keats further deadpanned that, like many other filmmakers, he had begun in pornography, although in his case it was pornography for plants, by which he meant a film of bees pollinating flowers. He further explained, with, one assumes, tongue temporarily out of cheek, "What I'm always doing is trying to pose thought experiments in the old-fashioned philosophical way . . . imagining from a radically different perspective circumstances that are very familiar to us, in order to make them unfamiliar and force us to start to pull them apart."[23]

It is telling that some of the most interesting thinking about the experience of radically different minds comes from a scientist whose experiments led well outside established scientific practice, and from a self-styled conceptual artist and experimental philosopher. We are near or past some sort of boundary here, and perhaps we should recognize that. We might also recognize that, with regard to extraterrestrial intelligence, science gets us only so far. Science historian Steven J. Dick puts it directly:

Science can as yet add nothing to the question of the physical, mental, and moral nature of intelligence beyond the Earth. At best, science may shed a pale light on the question of the possible physical forms of the alien, but it can say nothing about its mental evolution—much less about whether good or evil or some compromise of the two rules such intelligences as might exist in the universe.[24]

He continues, "For that, the speculations of science fiction . . . are as valid as anything science can suggest." We might add that science, likewise, has only so much to say about problems of communication and understanding between mutually alien intelligences. And these, too, are matters that science fiction writers have thought about a good deal. And so, with care and deliberation—and only for a single chapter—we cross the line.

CHAPTER EIGHT

Weird Life in Science Fiction

Extraterrestrial beings have appeared in science fiction for more than a century, and there are now enough to fill a large encyclopedia. In fact, there is such a compendium: *Barlowe's Guide to Extraterrestrials*.[1] Although its entries represent a vast and wide-ranging body of speculative organisms, only a few are of interest here, as only a few could be classified as weird life.* Probably the reason for their relatively small number is simply that authors find them unnecessary. Most science fiction authors use

---

* The definition of weird life that has been operative in much of this book—"life that is not descended from LUCA"—requires some modification here. Much extraterrestrial life in science fiction, given its age, location, and/or nature, cannot possibly be descended from LUCA, yet (rather improbably) is depicted as being indistinguishable from life we know. Of course, most renderings of extraterrestrial life in science fiction do not address questions of that life's ancestry to begin with. For the sake of convenience, in this chapter I'll disregard the issue of lineage and define weird life simply as life that is fundamentally different from that we know.

extraterrestrials to offer commentary on humankind—usually critical, occasionally flattering—and such aliens need have only superficial differences. Take, say, fourteenth-century samurai, give them purple skin and pointy ears, allow them to intone a few sage-sounding proverbs, and many readers, especially those who have fallen behind in their studies of feudal Japan, will be happy to believe they are getting a genuinely otherworldly take on the human condition.

It is true that many intelligent extraterrestrials in science fiction have a biology unlike that of humans. Ursula Le Guin's novel *The Left Hand of Darkness*, for instance, features beings that can change their gender. But even ambisexual life is not weird in the sense I mean here, and in fact in our nonfictional world it is fairly common. Many plants and animals, including a mollusk with the lovely English name slipper limpet (and the somewhat less lovely Latin name *Crepidula fornicata*), are male at certain times in their life cycle, female at others.

Some of the life depicted in science fiction that *is* weird in the sense I mean here is rendered with no explanation of its biology. The "Hooloovoo" from Douglas Adams's *The Hitchhiker's Guide to the Galaxy*, for instance, is "a superintelligent shade of the color blue."[2] No doubt a few graduate students somewhere spent a late night theorizing about its biochemistry, but most readers take it as the whimsical construction that Adams most probably intended, happily suspend their disbelief, and read on.

Suppose we define "weird life" broadly and generously, and begin with all the weird life in science fiction. When we remove depictions of alien life that is not truly weird (like Le Guin's ambisexuals), and remove impressionistic renderings (like Adams's Hooloovoo), we are left with a handful of attempts both to depict weird life *and* to explain it, accomplished with various degrees of plausibility and detail.

There are many depictions of life based on silicon. Among them are a planetwide ecosystem and a life-form with a biochemistry of silicon and superfluid helium—the latter organism suited to very cold environments. In fact, authors of science fiction have imagined several varieties of cryogenic weird life, including some that survive on Kuiper belt objects, the icy bodies that circle the Sun in orbits beyond Neptune and have surface temperatures a frosty 30 degrees above absolute zero. The Kuiper belt organisms, the size and shape of shrimp, have a biochemistry based on fluorocarbons with oxygen difluoride as the biosolvent; they warm themselves by secreting a pellet of uranium-235 inside their bodies and moderating its nuclear fission through their shells. A convincing depiction of life at thirty degrees above the point at which molecular motion ceases would be a formidable challenge for the most ingenious science fiction, but one writer envisioned still-colder organisms that survive by using a biosolvent of helium, which is liquid at just 2 or 3 degrees above absolute zero.

There is a minor science fiction tradition of life in stars.[3] Writers depicting such life face a rather different set of problems. The atmosphere of a Sun-like star has several layers, one of which is the "chromosphere." The temperature of a chromosphere reaches 1,000,000°C, hot enough to tear atoms and molecules apart, leaving their remnants as an ionized gas called "plasma"—hardly the stuff of which life is made. Nonetheless, several science fiction authors imagined beings that use magnetic fields to coax that plasma into chemistry and biochemistry. Again, such ideas would seem as strange as science fiction might manage. However, there are depictions of life taking forms and surviving in locales still more exotic—plasma-based organisms in the accretion disk of black holes, beings composed purely of quantum wave functions in the gravity wells of stars, and life in a parallel universe with different physical laws entirely.[*]

In most of these works the weird organisms are as shocked to learn of humans as humans are to learn of them. In many cases, the shock is only the story's beginning. When it has passed and both species have acknowledged the vast gulf separating their natures, they must somehow bridge that gulf, and to do it they must overcome enormous—in fact, unprecedented—challenges to communication and understanding. For the reader or viewer (many examples come from television and film), much of the pleasure of these stories derives from seeing humans and weird life cooperate to do exactly that. It is no surprise that there are many examples of such cooperation, and no surprise that one such example plays a role in what is probably the best-known encounter with weird life in all of science fiction.

That encounter appears in a screenplay by Gene L. Coon, a writer for the original *Star Trek* television series. The crew of the *Enterprise* encounters a silicon-based life-form. Mr. Spock

---

[*] The planetwide ecosystem based on silicon is featured in Alan Dean Foster's *Sentenced to Prism*. The organism based on silicon and superfluid helium appears in Arthur C. Clarke's "Crusade." The Kuiper belt organisms are featured in Robert L. Forward's *Camelot 30K*. The organisms that use liquid helium as a biosolvent are featured in Larry Niven's *Known Space* universe. The "flame-like inhabitants" of stars appear in Olaf Stapledon's *Last and First Men* and *The Flames*. Other stellar denizens make appearances in David Brin's *Sundiver* and Frederick Pohl's *The World at the End of Time*. The plasma-based life making its home in the accretion disk of a black hole is in Gregory Benford's *Eater*. Photino-based organisms and organisms composed of quantum wave functions are featured in several of Stephen Baxter's novels. Organisms that operate in a parallel universe appear in Isaac Asimov's *The Gods Themselves*.

(who, as most readers no doubt know, is an English-speaking extraterrestrial) initiates a telepathic interchange during which the creature learns enough vocabulary and calligraphy to etch the words "NO KILL I" in rock. You might think that its rendering of English letters in a perfectly executed Helvetica would earn some admiration from the crew of the *Enterprise*. Not quite. The apparent confusion of nominative and objective case gives them pause, and Captain Kirk wrestles with an interpretation. "A plea for us not to kill it?" he wonders, "or a promise it won't kill us?" Seeking an answer, Mr. Spock employs telepathy again and finds that they had underestimated the creature. Employing a poet's economy of language and a minimum of rock-eating acid, it had meant both. Soon mutual understanding on larger issues is achieved, Dr. McCoy dresses the creature's wound with "thermo-concrete," and on one planet at least, carbon-based and silicon-based life manage to coexist.

Since intelligent weird life of science fiction is weird in a variety of ways, it follows that humans and weird life who wish to communicate and understand each other must seek a variety of avenues for communication and understanding. A rather different avenue is suggested in the work of science fiction writer James White, who has said that the political unrest and violence of his native Northern Ireland compelled him to conceive of a place that represented its opposites. The setting for twelve of White's novels and numerous short stories is a space-based hospital, with patients that include delicate dragonfly-like chlorine breathers, ambulatory vegetables, and armor-plated elephantine beasts that can withstand atmospheric pressures four times that of Earth's at sea level. The facility itself boasts wards with various atmospheres and gravities. The attending physicians, themselves representing eighty-seven different species, are expected to understand exotic chemistries and treat equally exotic ailments—a challenge even to the most accomplished among them. They meet that challenge with "education tapes" that temporarily endow their user with the expertise of a physician who is a member of the species being treated. Owing to an imperfection in their design, the tapes convey not only the medical knowledge of the physician, but his, her, or its personality. Included in that personality are gastronomical and sexual appetites—for the stories' characters, more communication and understanding than is strictly necessary, and for the stories' readers, interesting and amusing plot complications.

Communication between humans and White's beings is challenging, among other reasons, because they experience pressures, chemistries, and radiation differently than we do. Another author has depicted beings who experience *time* differently than we do. As you might expect, communication between humans and these beings demands much effort and much imagination. As you also might expect, these beings inhabit a world radically unlike any described so far.

WHEN YOU LIVE UPON A STAR

A neutron star is the remnant of a supernova explosion. Far more dense than even a white dwarf, it is a mass of the Sun squeezed into a sphere the size of a city. While the electron shells surrounding the atomic nuclei of white dwarf stars are greatly compressed, the electron shells surrounding atomic nuclei in neutron stars are crushed entirely, so that the nuclei themselves are pushed together. In the 1970s, when pulsars were discovered to be rapidly rotating neutron stars, Frank Drake—the same Frank Drake who in 1959 conducted the first serious search for radio signals from other civilizations—gave popular lectures on the phenomenon. To help his audiences visualize the surface of a neutron star, he described it from the point of view of creatures who might be living there.

Seven years later, a physicist and engineer named Robert Forward modified the same idea as a basis for a science fiction novel called *Dragon's Egg*. Forward imagined tiny creatures about the size of a sesame seed, composed, like the star's crust, of densely packed atomic nuclei and having a metabolism that relied on nuclear chemistry. Familiar chemistry depends on the electromagnetic force, but Forward's hypothetical nuclear chemistry was mediated by the strong force, and because the force is so much stronger than the electromagnetic force, the reactions of the nuclear chemistry are much faster—in fact, about a million times faster. The effective relative timescale between Forward's beings and humans would be a million to one, meaning that one of the beings would live its natural life span, the equivalent of our three score and ten, in all of fifteen minutes. Although both species manage to send and receive messages with radio, this temporal disparity presents an obvious challenge. It is met because the inhabitants of the neutron star have the patience to wait half their lives to complete a single exchange with the humans, and because the humans (who have a somewhat easier time of it) possess the technology to greatly slow recordings of the messages they receive. All in all, organisms living on the crust of a neutron star was a fairly wild idea, but Forward's rationale was sufficiently grounded that he could later publish it in the journal *New Scientist*.[4]

The idea of weird life on neutron stars is remarkable, among other reasons, because such life would be incredibly dense, having adapted itself to places where matter is greatly compacted. Another sort of weird life, adapted to regions where matter is spread thinly, might be incredibly diffuse.

NEBULAE

Sir Fred Hoyle may be best known as the astronomer who coined the phrase "big bang" (derisively, as it happens) to describe the currently most favored theory for the universe's origin. Hoyle also enjoyed an avocation as an author of science fiction. In 1957 he published a novel called *The Black Cloud*. Its central character, and the cloud of the title, is a nebula of hydrogen and more complex molecules that, although diffuse, is organized in the manner of a living being. (Like Forward's neutron star life, Hoyle's cloud was well grounded in science—enough that Freeman Dyson, in his hypothesis of life in a distant future, cited Hoyle's novel as inspiration.[5]) The cloud propels itself by manipulating magnetic fields, and thinks with electrically charged dust particles—the rough equivalent of a vast set of neurotransmitters. When a group of radio astronomers realize that the cloud is both alive and intelligent, they manage to communicate with it via radio waves, and translate its messages into audible English speech.

The cloud catches on quickly and, offering a nice reversal of perspective, politely explains to the humans their sheer improbability: "Your first transmission came as a surprise, for it is most unusual to find animals with technical skills inhabiting planets, which are in the nature of extreme outposts of life."[6] The cloud notes that the gravity on a planet's surface severely constrains the size of animals and so the size—and therefore the complexity—of their brains. Moreover, gravity forces the development of "muscular structures" to enable mobility and "protective armor" like

A griffin. The griffin has the head, talons, and wings of an eagle and the body of a lion. According to legend, griffins mated for life; if either partner died, the other would continue throughout the rest of its life alone, never to seek a new mate. This drawing (by John Tenniel) is of the griffin featured in Lewis Carroll's *Alice's Adventures in Wonderland*.

Riftia pachyptila. *Riftia* tube worms colonize hydrothermal vent habitats between broken pieces of lava. (Courtesy NOAA *Okeanos Explorer* Program)

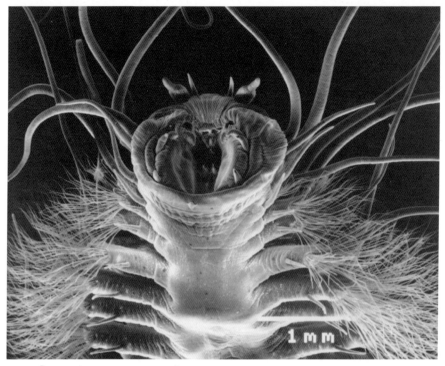

Methane ice worm (*Hesiocaeca methanicola*). Scientists discovered this organism in 1997 living on and within mounds of methane ice on the floor of the Gulf of Mexico. (Courtesy NOAA)

Mono Lake, California. The site of a discovery (now largely discredited) of life that is reputed to have replaced some of its phosphorus with arsenic. (Courtesy NASA)

Titan seen through Saturn's rings. The haziness of Titan's atmosphere is obvious in this image, in which the moon appears behind two of Saturn's rings. Epimetheus, another of Saturn's sixty-two moons, appears just above the rings. (Courtesy NASA/JPL/Space Science Institute)

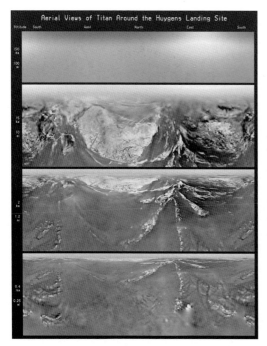

Approaching Titan. Views from the descent/imager spectral radiometer on the European Space Agency's *Huygens* probe at four altitudes. (Courtesy ESA/NASA/JPL/University of Arizona)

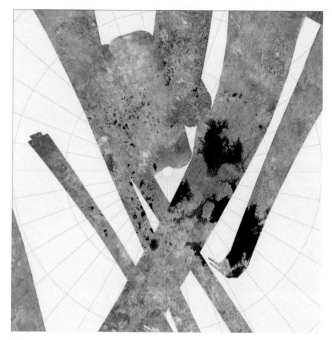

Hydrocarbon lakes in Titan's north polar region. The dark areas are believed to be large bodies of liquid hydrocarbons. (Courtesy NASA/JPL/USGS)

Neptune's moon Triton, site of a hypothetical cryogenic biochemistry. Visible in this image are relatively fresh nitrogen frost deposits, "cantaloupe" terrain, and landscapes apparently produced by liquids erupting from the moon's interior and freezing on the surface. (Courtesy NASA/JPL/USGS)

Floaters. An imaginative depiction of "floaters" in Jupiter's stratosphere. (Courtesy Dan Durda, Fellow, International Association of Astronomical Artists [IAAA])

Life along the shore. An imaginative depiction of life along the shoreline of a lake on an extrasolar planet. (Courtesy Dan Durda, Fellow, IAAA)

An alien sky. An imaginative depiction of avian life over the sea of a giant planet's moon. (Courtesy Dan Durda, Fellow, IAAA)

A surprise. An imaginative (and whimsical) depiction of one of NASA's *Mars Exploration Rovers* approaching an intriguing find. (Courtesy Dan Durda, Fellow, IAAA)

skulls to shield the brain. Naturally, these requirements exact a cost, constricting the size of the brain. Since the cloud is unencumbered by gravity, it labors under no such restraints. It enjoys another advantage over planet-bound organisms. Like most life on Earth, the cloud derives energy from sunlight to initiate and sustain chemical reactions. But because its surface area is enormous, it can avail itself of far more energy than can all of Earth's biomass together. The cloud concludes its lesson to the humans with barely concealed condescension, noting, "By and large, one only expects intelligent life to exist in a diffuse gaseous medium, not on planets at all."[7]

A RANGE OF WEIRD LIFE

Most of the authors who depicted weird life did so for only one or two works. In fact, only a few science fiction authors have produced large bodies of work focusing on weird life. Of these, the most prolific may have been Henry Clement Stubbs, writing under the pen name Hal Clement.

Clement studied astronomy at Harvard and taught high school science, and all Clement's aliens are the sort of intelligent weird life one might expect from the mind of a good science teacher—one who could look at a chemistry experiment and imagine a world inside it.[*] One of Clement's novels is narrated by

---

[*] To depict weird life in general terms, an author doesn't need to invent a different biology or biochemistry—welcome news to that author, as such efforts would require specialized knowledge. Probably we shouldn't be surprised, though, that many of these stories are by authors with training in the sciences. Fred Hoyle, who imagined a sentient interstellar nebula, was a professional astronomer. Robert Forward, who described beings with a nuclear biochemistry, was a physicist. Stanley G. Weinbaum, whose 1934 short story "A Mar-

a being that breathes sulfur and drinks copper chloride. Another features life on a planet with an atmospheric pressure 800 times greater than Earth's. Yet another depicts an alien and a human, with radically different conceptions of normal, making their way across a planet whose conditions are at the absolute tolerable limits for each, and gradually coming to mutual understanding. It is the plot of a hundred Hollywood buddy movies, except that the divide that separates our heroes is not merely cultural or economic or racial. It is biochemical.

Clement's best-known work is a novel called *A Matter of Gravity*, its action set on a massive (and, of course, fictional) planet whose rapid rotation causes it to bulge outward at the equator and makes for gravity that increases from a challenging 3 times Earth gravity at the equator to a crushing 300 times Earth gravity at the poles. The intelligent species on the planet are a kind of armor-plated centipede half a meter long, their technology at a stage roughly comparable to that of medieval Europe. The centipedes have a fear of heights—rational, as fears go, in that in the planet's gravity things fall even short distances at the speed of a rifle bullet. Otherwise, they think and behave like any sampling of intelligent humans with a scientific bent, making and testing hypotheses, and building on the results. When they meet the arrivistes from Earth, the centipedes bargain for knowledge. They are quick studies of human psychology, and like the hayseed in the city, they get the better end of the deal—enough

---

tian Odyssey" features a silicon-based creature that creaks when it moves and produces silicon dioxide (not carbon dioxide) as waste, had a degree in chemical engineering. Gregory Benford, who described life near black holes (among other exotic locales), is a practicing astrophysicist.

that they can bypass a scientific revolution or two. All of which is to say that the centipedes are rather familiar character types. Their weirdness is only skin—or, armor-plate—deep.

All of Clement's aliens are rational thinkers who think with scientific method, or learn to. All hold that the universe is comprehensible. The resulting view of nature—much the same in Hoyle's *The Black Cloud* and Forward's *Dragon's Egg*—is reassuring. The universe may be fascinating and awe inspiring. It may hold wonders we have not imagined. But all of it, with enough time and effort, can be understood.

There is, though, another sort of weird life in science fiction, and it is more disturbing, in part because it is truly and deeply weird—weird, we might say, all the way down. One of the best examples is one of the earliest, in David Lindsay's 1920 novel *A Voyage to Arcturus*. The planet on which the action takes place has not only weird life, but an alternative to Darwinian natural selection, a hyperactive and directed sort of Lamarckian evolution in which creatures can actually will the properties of their progeny, the result being a natural world so "energetic and lawless" that no two creatures are alike.[8]

A more recent example appears in Stanislaw Lem's 1961 novel *Solaris*, with film versions in 1972 and 2002. The novel and films differ slightly, but in each, scientists from Earth have discovered a planet that is covered by what they describe (most easily but probably inaccurately) as a living ocean. In the generations that follow, thousands upon thousands of scientists study the ocean, venture hypotheses, and gradually construct theories as to its internal ecologies, its biochemistry, and its origin. Some believe it is conscious; others suspect it is merely very complex. On all these matters there are many schools of thought, dissenters within each, and naturally, decades of scholarship that fill whole libraries. Yet for all of this, not one of the scientists can say with any assurance exactly what it is they are studying.

So we have, in science fiction, a range of weird life. At one extreme the centipedes—superficially weird but fundamentally familiar; at the other the living ocean—profoundly weird and utterly unknowable.

Is there, as it were, a weird medium? In fact, a large body of work depicts life that human characters cannot hope to understand intellectually, but can appreciate aesthetically. These organisms appear in the fiction of Joseph Henri Boex, a Belgian expatriate who in the 1880s settled in Paris and adopted a pen name that, Englished, is "Rosny the Elder." (He and a younger brother who had coauthored many works parted professional ways and split the name they had been using—J. H. Rosny— according to birth order.) The name may sound as though it belongs to a medieval biographer of saints' lives, but as we'll see, Rosny's concerns strayed far from the doctrine of most churches. Only a few of Rosny's works were ever translated from the original French, and most of those are out of print, probably because critics have been less than charmed by what one called his "prolixity, maudlin sentimentality, and awkward stylistic mannerisms."[9] Such qualities seem not to have troubled Rosny's readers: in his own lifetime he enjoyed a popular following that in size rivaled those of Jules Verne and H. G. Wells.

Rosny's plots were kaleidoscopic mash-ups of, well, everything. One features a Stone Age tribe discovered living in the Arctic alongside woolly mammoths, "rescued" by an explorer and transplanted to northern Africa. Another concerns a society of intelligent vampire bats. Still another involves a young man who can see beings living invisibly alongside us, sometimes brushing against us, and (shades of a shadow biosphere) subtly impinging on our world as we impinge on theirs.

Like Wells, Rosny took inspiration from Darwin's theory of natural selection, and was especially intrigued by its claim that nature had produced and would continue to produce "endless forms most beautiful and most wonderful."[10] He imagined many: living minerals that threatened prehistoric tribes of humans; "ferromagnetic" entities that, far in Earth's future, supplant organic life; and on Mars, luminous networks of intelligent phosphorescence. All are so alien that humans cannot understand them or communicate with them, but not for lack of trying. Rosny's humans are adventuresome sorts. In one story, a male human scientist meets a female Martian who, like all her species, is possessed of "trilateral symmetry"—that is, six eyes, six ears—and before you can shrug *"l'amour,"* she and the scientist are deeply involved in matters not strictly scientific. Although the Martian *maîtresse* may not be not weird life in the sense I've been meaning, the open-mindedness that allows Rosny's scientist to see the beauty in her trilateral symmetry also allows him to be properly awed by organisms that are quite beyond human understanding.

We've discussed hypotheses of life that differs from the familiar in many ways—its chemical structures and pathways and the "handedness" of its DNA. We've discussed hypotheses of life that uses a liquid medium other than water, life that uses mediums that are not liquid, and life that uses no medium whatsoever. We've discussed hypotheses of life driven by a nuclear chemistry. And we've considered the places such organisms might call home: rock surfaces in the American Southwest, hydrothermal vents on the ocean floor, Martian permafrost, the water-ammonia oceans of Jupiter's moons, the cold methane and ethane lakes on Titan, the deep hydrogen-rich atmospheres of giant planets, the exotic ices in comets, the crusts of neutron stars, and the vast reaches of space itself. We would seem to have amassed as complete an inventory of weird life and weird-life environs as is possible.

Yet some scientists think there may be life weirder still. All the previous—from the microbes that might produce desert varnish, to the hydrogen-eating dirigibles, to the sentient interstellar nebulae—if they exist—would abide by the same natural laws that you and I abide by: the strength of gravity, the particular mass of the subatomic particles of which we are composed, and so forth. These laws are so fundamental that few of us have bothered to think of them, much less wonder what worlds might result if they were otherwise. Yet there may be places where such laws *are* otherwise, and a handful of scientists suspect that at least a few such places might harbor life—the weirdest life of all.

CHAPTER NINE

Weird Life in the Multiverse

P ossibilities unrealized and roads not taken exert a strong pull on our imaginations, and many of us surrender to that pull annually, to reread Dickens's *A Christmas Carol* and watch Capra's *It's a Wonderful Life* one more time. We are drawn to both for their what-ifs and what-might-have-beens, but we leave them thinking about our own—another year's worth every year. There is something both wondrous and a little terrifying in such thoughts, and it seems fitting that in the story and film they are premised on the supernatural, introduced by a ghost and an angel. But as guides to alternate realities, it turns out that ghosts and angels aren't necessary. In fact, findings in science— especially in theoretical physics and cosmology—are the basis for many more recent fictional alternate realities. Such places are called, among other things, parallel universes.

Parallel universes have provided grist for innumerable comic books and a whole subcategory of science fiction. In the last decade or so, when science fiction broke into the cultural mainstream, parallel universes came with them. Nowadays they seem to be everywhere. They supply plots for Hollywood and television techno-thrillers, for cerebral independent films, and for dense philosophical novels whose characters struggle with questions of fate and free will. One reason this is happening now, no doubt, is the news that all the what-ifs and what-might-have-beens may actually be out there somewhere.

In recent years, cosmologists and theoretical physicists have devoted much attention to the concept of a *multiverse*—that is, a set of universes whose number is at least very large and may be infinite. It's an unsettling idea, in that it challenges not only our intuition but our vocabulary. Conventionally, the word "universe" has meant "everything" or "all there is." And so we might reasonably ask how anything can be outside the universe or apart from it, or indeed how there can be more than one. By way of an answer, we'll need to back up a bit.

THE TRADITIONAL UNIVERSE

In the sixteenth century the Italian philosopher Giordano Bruno asserted that nature's perfection and God's power necessitated an infinity of worlds.[1] Several hundred years hence, and by a rather different chain of reasoning, scientists arrived at much the same conclusion. By the twentieth century, astronomers and cosmologists had amassed evidence suggesting that the universe—all of space and everything in it—extended outward to infinity in all directions. (The alternative seemed downright nonsensical: how, after all, could there be an end to space?) They had also amassed evidence suggesting that the laws of nature were the same throughout. The upshot—a universe that was infinite and more or less homogenous—became the traditional model of the universe, and like most traditional models, it had variants.

One variant, introduced by Albert Einstein, resulted from the possibility that space might be curved. If three-dimensional space were curved as a two-dimensional surface can be curved on, say, a sphere, then the consequence would be a universe that was finite (there is only so much surface on a sphere) yet unbounded in that (as on the surface of that sphere) if you traveled far enough in a straight line, you would arrive back where you started—from the opposite direction. Of course, such a universe might be curved in other ways: like the surface of a saddle whose edges extend to infinity, making for a universe that was unbounded and infinite; or like the surface of a pretzel, making for a universe that, like the surface of the sphere but far more interesting, was unbounded but finite. At present, the variant is largely discredited. Cosmologists have little evidence to suggest that the universe on large scales is curved and—from maps of the cosmic microwave background, that relic radiation from the big bang—persuasive evidence that it is flat.

A second variant of the traditional model was the island universe. It held a finite amount of matter in an infinite space. Galaxies, our own among them, were huddled together in a large collection that thinned out at its edges and gradually gave way to an emptiness that extended to infinity. At present, however, astronomers have detected no "thinning out." As far as their instruments can probe—nearly 42 billion light years out—they have found galaxies and stars made of the same stuff as galaxies and stars nearby, and behaving according to the same laws of gravity and motion. They see structures on the largest scales, bubbles of mostly empty space 300 million light years across with skins made of clusters and superclusters of galaxies, spread evenly, also as far. They have every reason to expect that such structures continue beyond the limits of their vision.

So we are returned to the traditional model in its original form—a universe that is infinite, flat, and populated with galaxies and stars everywhere. It has many merits. It is the model of the universe with historical precedent and the model in agreement with observation. It is also the model in accord with what is called the "cosmological principle" or, more commonly, the "principle of mediocrity"—the working assumption of cosmologists, expanded and codified from Copernicus, that we are nowhere special.* Finally, the traditional model of the universe is also the default model—that is, the model used by cosmologists for most calculations and simulations. Yet for all this, it carries with it some rather bizarre implications.

QUANTUM STATES

Consider the following. First, only a finite number of particles can fit into a given volume of space before it collapses of its own weight into a black hole. Second, at any given moment, each of those particles can have only a finite number of positions and speeds, and all the particles in a given volume can be arranged in only so many ways.† Third, there is an infinite amount of space

---

* The terms "Copernican principle," "cosmological principle," and "principle of mediocrity" have often been used interchangeably, although the last, as defined by physicist Alexander Vilenkin, has a specific valence. It asserts that since we are typical of intelligent beings in the multiverse, our observations should be typical of those made by all such intelligent beings. (Greene, *Hidden Reality*, 180) Since Vilenkin's definition most suits our purposes, I will use "principle of mediocrity" throughout.

† The world of quantum physics has aspects that are counterintuitive, one of which is its intrinsic uncertainty. A subatomic particle like an electron, for instance, cannot have a position and speed simultaneously; it is for this reason that a quantum physicist speaks not of a particle's position and speed but

and an infinite number of volumes of that space. From these three straightforward and reasonable-sounding observations follows a rather startling conclusion: all possible arrangements of particles must be out there somewhere, and more mind-boggling still, all possible arrangements of particles must occur not just once, but an infinite number of times.

In case you weren't taking the principle of mediocrity personally, now you have a reason. Because one such arrangement is you. You have an infinite number of doppelgängers, the nearest of which is at a distance so great as to be unimaginable yet, strange to say, measurable. Physicist Max Tegmark, by calculating the number of particle arrangements in an appropriately sized volume of space and assuming that arrangements are distributed randomly, estimates that he or she (your doppelgänger, that is) is 10 to the power of 10^{29} (that is, $10^{10^{29}}$) meters away. It's quite a distance. By way of comparison, the radius of our observable universe is far smaller—a mere 4×10^{26} meters.

There is another consequence here, rather nearer our interests. Because all possible particle arrangements are realized somewhere, it follows that the weird life described in previous chapters must exist. Unless they violate some natural law, then no matter how improbable their biochemistry, the arsenic eaters, the ammonia drinkers, and the living dirigibles are also out there somewhere. And as with your doppelgängers, there must be an infinite number of each. The only qualification would concern their distribution. The less probable an organism, the fewer there would be in a given volume of space.

---

of its "quantum state." Likewise, an arrangement of particles within a volume would have an observably distinct quantum state. For our purposes the word "arrangement" will suffice.

MULTIVERSES

This model of the universe, with doppelgängers, familiar life, weird life, and much, much else repeated through infinity, also contains an infinite number of observable universes, each overlapping others, each 84 billion light years across and growing outward in all directions one light year every year. Of course, in the sense that the space between them is continuous, they are not separate universes at all, but merely regions within one universe. When most cosmologists speak of alternate universes, parallel universes, and a multiverse, they have something else in mind: whole sets of universes, perhaps an infinite number, each as real and—strange to say—as infinite as the traditional model.

On introduction, such an idea seems a bit untethered, and laypersons may be forgiven for suspecting it to be a late-night notion propped up with a few hypotheses hastily cobbled together the morning after. In fact, though, cosmologists and theoretical physicists have not been looking for multiverses. Quite the contrary: for reasons we'll discuss shortly, many would prefer there were no such thing. But as string theorist and science writer Brian Greene notes, multiverses are turning out to be "harder to avoid than they are to find."[2]

THE QUANTUM MULTIVERSE

Quantum mechanics is the theory describing the laws of physics that explain the universe on very small scales and underlie it on large scales. As theories go, it has been phenomenally successful—explaining the structure of atoms, radioactivity, superconductivity, the effects of electrical and magnetic fields, and the thermal and electrical properties of solids. It has also made possible the technologies of lasers, transistors, and electron microscopes. Since its beginnings more than half a century ago, not a single experiment has contradicted the predictions of quantum mechanics.* Yet peel away the skin of those predictions and you'll find a mystery. Quantum physicists make testable predictions with a "quantum algorithm," and they do not agree on its meaning; that is, they do not agree on exactly *why* the algorithm works. Thus, quantum mechanics comes with different approaches, all of which are attempts to explain the algorithm's meaning and the uncertainty associated with a particle like an electron.

The "many-worlds" approach, put forth in 1957 by then Princeton graduate student Hugh Everett III, states that subatomic activity continually creates new universes—or, as more recent proponents would have it, continually differentiates

---

* Alas, a common misunderstanding of scientific practice and terminology still necessitates a footnote. The oft-heard dismissal "it's just a theory" fails to recognize the full meaning of the word "theory." A real scientific theory is a self-consistent set of hypotheses that make predictions about nature that are testable and falsifiable. When the prediction of a certain hypothesis is shown to be correct (as when, for instance, a hypothesis of Einstein's theory of general relativity predicted that sufficiently precise measurements would show that the Sun's gravity was bending starlight), the theory gains credibility and authority. As more and more of a theory's hypotheses are shown to be true, the theory is judged successful. But the theory in its entirety is not proved, and probably never will be. A theory is always provisional simply because we most likely will never know everything, and some things we do not know may someday disprove a certain hypothesis, thus collapsing the theory it helped to support and sending the hypothesizer back to the proverbial drawing board or proverbial cocktail napkin. A theory's provisional nature, far from being a flaw, is essential to scientific advancement.

among identical copies of existing universes. These universes exist in what theoretical physicists call "another quantum branch in infinite-dimensional Hilbert space" and what science fiction authors call "another timeline." At first, Everett's work was neglected; in the late 1960s, however, physicist Bryce DeWitt dusted it off and presented it to larger audiences, emphasizing the many-worlds aspect. The approach might have been dismissed outright, but for the fact that it was competing against a weak field. Nobel Laureate physicist Steven Weinberg called it "a miserable idea except for all the other ideas."[3] For several decades a few theoretical physicists speculated about other universes, but most put all approaches aside and adopted an attitude that some called "shut up and calculate" and others (no doubt in need of a more respectable-sounding phrase) called "pragmatic instrumentalism."

Then, in the late 1970s, an Australian cosmologist and theoretical physicist named Brandon Carter turned his own instrumentalism to questions that were anything but pragmatic. He wondered how the universe might have been different had the laws of physics been other than what they are. He noted, as had several before him,* that if those laws had been much different, and in many cases only a little different, the universe would not have been able to support complex chemistry, let alone biochemistry or biology—or us.

---

* In 1904, British naturalist Alfred Russel Wallace observed, "Such a vast and complex universe as that which we know exists around us, may have been absolutely required . . . in order to produce a world that should be precisely adapted in every detail for the orderly development of life culminating in man." (Wallace, *Man's Place*, 256–7)

THE STANDARD MODEL OF PARTICLE PHYSICS

Physicists' best understanding of these laws is expressed in the "Standard Model" of particles and forces. The Standard Model states that everything in the universe is made from twelve fundamental particles—six types of quarks (which compose protons and neutrons) and six types of leptons (the most familiar of which are electrons and neutrinos). The model further states that these particles are governed by three fundamental forces: the "electromagnetic force," the "strong nuclear force" (or simply the "strong force"), and the "weak force." The model does not have a place for gravity, nor does it explain many other features of our universe; nonetheless, it is regarded as successful in that it explained a variety of experimental results and predicted the existence of several subatomic particles years before their actual discovery.

Let's take a moment to review the Standard Model's details. Two types of quarks go to making protons and neutrons, and protons and neutrons go to making atomic nuclei. Just as protons and neutrons have particular masses, so do the quarks that compose them. The quarks (and their masses) are governed in various ways by the fundamental forces. The strong force holds quarks together to make protons and neutrons, the weak force causes a type of radioactive decay, and the electromagnetic force governs electricity, magnetism, and other electromagnetic waves.

What puzzled Carter and others about the masses of quarks and the strengths of the forces is that all seem finely tuned to a universe that allows life. Consider the quarks. Adjust their masses ever so slightly—so that, for instance, neutrons become 2 percent heavier than protons—and you couldn't have stable oxygen or carbon. No oxygen and no carbon would mean no life. Adjust the masses of quarks so that protons are much heavier than neutrons, and you couldn't have stable hydrogen. No hydrogen would mean no life. For the fundamental forces the situation is much the same. Changes to the strength of any of them, even small changes, would make a universe inhospitable. If the electromagnetic force were a bit stronger, atoms would not share electrons and chemistry would be impossible; if it were a bit weaker, atoms could not hold on to electrons, the universe would be populated only by loose subatomic particles, and chemistry (again) would be impossible. If the strong nuclear force were stronger, stars would turn all their hydrogen into helium and then iron, and we'd have a universe without hydrogen. If it were weaker, complex atomic nuclei couldn't form, and we'd have a universe without carbon. If the weak force were stronger, atomic nuclei would decay before heavy elements could form; if it were weaker, then (as with a stronger strong force), all hydrogen would be turned into helium.

The strangeness doesn't end there. The universe we know has features not explained by the Standard Model—features that physicists and cosmologists call "cosmic parameters"—and these also have values that are quite specific and seem finely tuned to allow life. They also seem, for lack of a better word, *arbitrary*. Take, for instance, the strength of gravity. It has a specific value expressed by the gravitational constant: $G = 6.67 \times 10^{-11}$ cubic meters per kilogram per second squared. This equation, one can't help noting, is so lacking in elegance that it might have been generated randomly. If you guessed that it's not the prediction of a theory, you'd be right; the equation representing the strength of gravity had to be found by direct experiment. But for all its coarse appearance, that equation is precise, and for that we should be thankful. Were gravity much stronger, cosmic expansion would have slowed, halted, and reversed itself, and the universe would have collapsed almost before it had a chance to start. Were it much weaker, the material created in the big bang would have continued to expand so quickly that it would have dissipated, and we'd have a universe of very diffuse particles.

Naturally, this question arises: Why should the values of the particles and forces be what they are? To this Brandon Carter had an answer. If there are a great number of universes, and laws vary from universe to universe, there may be no reason, and to look for one would be pointless. It is to be expected that we find ourselves in a universe with laws conducive to our existence; obviously it couldn't be otherwise. If anything explained these laws, Carter said, it was what statisticians call a "selection effect"—something that, on this largest and most fundamental of questions, scientists were failing to take into account.

Carter presented the idea in 1973 at a conference commemorating Copernicus. It was a rather pointed selection effect on his part, in that it ran counter to the Copernican-inspired principle of mediocrity and, rather more immediately, counter to an extension of the principle proposed by Fred Hoyle (the same Fred Hoyle who some years earlier had imagined a sentient interstellar cloud). As you may recall, Hoyle was a proponent of the "steady state" theory of cosmology, which postulated that the expanding universe was never in a state of higher density, that there was no big bang, and that matter is constantly being created out of empty space. Now Hoyle was suggesting that the principle of mediocrity should apply not only to space, but to time. In other words, as we assume that the universe is the same everywhere in space, we should assume that it is the same throughout time; and as we should expect to find ourselves at no special place in the universe's space, we should expect to find ourselves at no special place in its history. In Carter's view, this was one expectation too far, particularly in light of recent findings concerning the nature of the very early universe that, all evidence suggested, was a very different place. Carter was suggesting that perhaps we were someplace special after all. He called his idea the "anthropic principle," expressed formally (and in his words) as, "What we can expect to observe must be restricted by the conditions necessary for our presence as observers."[4]

In a very short time the anthropic principle generated hundreds of papers, several books, and several versions of the principle, the most audacious being the "participatory" form, posited in 1986 by theoretical physicists John D. Barrow and Frank Tipler, which asserted that the laws of physics and the universe are destined to produce observers of those laws and that universe. In other words, it suggested that life and intelligence, in some indeterminate manner, brought the universe into being.

By comparison with Barrow and Tipler's rendering, Carter's original version was tame, but it was controversial nonetheless, and remains so to this day, for two reasons. Recall that the Standard Model is regarded as incomplete and provisional. Many theoretical physicists believe that the laws of physics can be explained—and someday will be explained—as consequences of a single underlying principle, a synthesis of general relativity and quantum mechanics called quantum gravity. In something of an overstatement, it's also called a "theory of everything"[*]— although it probably won't explain, for instance, why fools fall in love or why there's always water in my basement. Nonetheless,

---

[*] Accordingly, the Standard Model is sometimes termed the "theory of almost everything."

many have devoted their careers in the attempt to discover that theory, and have regarded talk of anthropic rationales and alternate universes, with its implications that their work was a waste of time, as waving the flag of surrender. There is a related reason. Although Carter's anthropic principle constrained conditions to those we observe, it did not explain those conditions, and there was a danger that constraining would be confused with explaining. There was a danger, in other words, that cosmologists would come to believe they had explained a natural phenomenon like the strength of gravity merely because they had shown it to be compatible with life.

THE INFLATIONARY MULTIVERSE

In the mid-1990s, astronomers and cosmologists were amassing evidence to support a set of hypotheses that yield a model of the "inflationary" multiverse. The model implied that shortly after the big bang, the universe underwent a very rapid expansion, with some parts of space inflating like quickly rising bread even as they produced spaces within them (like pockets of air within that bread) that, once formed, stretched far less rapidly. The resulting picture is of a fantastically enormous expanse of ever-inflating space, doubling in size every 10^{-34} second or faster, and all the while producing ever-greater numbers of pockets of space. Each of the pockets, it should be said, is its own universe. And one of them is ours. For reasons explained by mathematics rather outside our interests here, while finite when viewed from its outside, each universe (ours included) would be infinite when seen from within.[5]

The inflationary multiverse, many think, may be the stage on which Carter's anthropic principle is realized. Cosmologists believe that as the very young universe expanded and cooled to a more stable state—more precisely, a "metastable vacuum state"—things gelled. It was a bit like a game of spin the bottle. As long as the bottle spins, it is unstable. It is stable—or at least more stable—only when it stops spinning. But exactly where it stops spinning is determined by many factors: an unevenness of the floor, air turbulence, and so forth—all the products of chance. Likewise, the universe's initial conditions—the densities and motions of matter—were products of chance, created early in its history by quantum fluctuations. Because the fluctuations were perfectly random, they were able to produce all possible metastable vacuum states. Consequently, so this model suggests, for each pocket universe, with its own big bang and its own expanding and cooling, the metastable vacuum state would be different. Therefore, physical constants like ratios of the masses of subatomic particles might be different, strengths of fundamental forces might be different, and perhaps most strangely, *dimensionality* might be different, and it might different in any number of ways—including the total number of dimensions, the parts of the total number that are compacted and unseen, and the geometry and topology of each.

Exactly how many metastable vacuum states (and pocket universes) might there be? No one really knows, but string theory—a leading contender for the theory of everything—allows an estimate. One prediction of string theory is that what we termed the traditional universe resides within a kind of substrate of space that physicists call a "brane." More or less as a page in a closed book is near other pages yet slightly offset from them, our brane is near other branes (with their attendant universes), yet slightly offset from them. In a hypothetical string theory "landscape," there are as many as 10^{500} different unstable vacuum states, meaning 10^{500} various sets of constants, particles, and dimensionalities, and 10^{500} directions for the spinning bottle, when it stops spinning, to point.

The environments of most other pocket universes are likely to be unfriendly to life, and you probably wouldn't want to visit one. You couldn't even if you did want to. Unlike the observable universes within the traditional universe, pocket universes are truly separate, and growing more separate by the second. They are driven apart with a speed proportional to the ever-growing space between them, such that two pocket universes sufficiently distant from each other are moving away from each other faster than the speed of light. If you started from one pocket universe, you could never reach the other, no matter how fast you traveled and no matter how much time you had. For these reasons you might think that we must abandon all hope of learning anything about them, and you might be right. But many physicists have reasons to suspect otherwise.

First, there is a precedent. The space beneath the event horizons of black holes is a place we will likely never see or visit. Nonetheless, theoretical physicists routinely use Einstein's theory of general relativity to describe the nature of that space, and they do so with some confidence. Second, evidence that physical constants have changed in our own universe—even a little bit— would show that they *can* change, and lend support to ideas of universes where they might be very different. (Intriguingly, there is evidence, albeit controversial evidence, of such a change. In 2001 a team of physicists reported observations of spectral lines produced by very distant quasars suggesting that 6 billion years ago, the electromagnetic force [specifically, the fine-structure constant characterizing the strength of the electromagnetic interaction] was slightly weaker.)[6] Finally, there is the possibility of experiment. If, sometime in the future, the hypotheses of string theory can be tested experimentally, then its prediction of pocket universes will gain authority.

There is, however, a way that the existence of a multiverse might be tested now: by predicting what scientists call a probability distribution. The thinking is as follows. Suppose the speculation of the previous pages is indeed the case—in other words, that our universe is one of an incredibly large set of universes, each with different physical constants, strengths for the fundamental forces, and dimensionalities. There is no reason to suppose that our universe is the only one that allows life; it is rather more likely that it is one among a whole subset of universes that allow life. If it is, then context changes everything. The only violinist in a high school concert orchestra who matriculates at Juilliard and suddenly finds herself in a roomful of violinists will learn that she is nothing special. Likewise, within a subset of universes that allow life, we would learn that ours is nothing special. In fact, within the subset of Juilliard violinists and life-allowing universes, the principle of mediocrity returns in full force. If our violinist is merely typical of violinists matriculating at Juilliard, odds are she won't be chosen to play first chair. Likewise, if our universe is merely typical of universes that allow life, odds are it is not among the few universes whose conditions for life are optimal. Rather, odds are it is among the many that meet those conditions with the slimmest of margins. An illustration may help.

Imagine a dartboard affixed to a wall, and if you like, imagine the wall extending a great distance in all directions. Suppose that the bull's-eye in the dartboard represents the optimal values of the constants, force strengths, and dimensionality for a universe that allows life. There are several rings around the bull's-eye, and they are of equal width, meaning that the larger the ring is, the greater is its surface area. Suppose that the smallest ring represents values slightly less optimal for a universe that allows life, the ring around it represents values less optimal still, and so on to the largest, outermost ring, which represents values that just barely meet the conditions necessary for a universe to allow life. All places on the wall off the dartboard, of course, represent values that fail to meet those conditions.

Now suppose that we throw darts at the wall, that they strike in a perfectly random manner, and that every place a dart strikes, a universe is created. Soon darts cover the wall and the dartboard evenly. The bull's-eye, having the smallest surface area, contains the fewest darts, the ring around it somewhat more, the ring around it still more, and so on. Because the outermost ring presents the most surface area, we may expect that there is a greater chance that a dart will hit it. Upon examination, we find that indeed, this outer ring contains more darts than any other ring and, of course, a great many more darts than the bull's-eye.

If our universe is typical of those with the conditions necessary to allow life, there is a greater chance that it is represented in the outermost ring than in any other. In other words, in our universe the conditions necessary to allow life would be far from optimum; in fact, the values of the constants, masses of fundamental particles, force strengths, and dimensionality for a universe would meet the necessary conditions just barely. So, we are now in a position to ask, *do* the values meet the conditions just barely? In the case of many values, the answer is yes—and in one case, a decided yes.

THE COSMOLOGICAL CONSTANT

Space many of us think of as empty is actually teeming with virtual particles that generate a repulsive force—the dark energy that drives galaxies apart at an accelerating rate. The number representing the value for the strength of that force is called the *cosmological constant*. For the first part of the twentieth century, no one knew what that number was. Then, in the 1970s, theoretical physicists calculated how much dark energy resides in a given volume of space, and they predicted a number for the constant. Their prediction surprised them: the number represented so much energy that galaxies would never have had a chance to form, and they assumed they had made a mistake. But upon careful review they concluded that there was no mistake—and yet the galaxies, obviously, had formed. Astrophysicists and cosmologists suspected that something was neutralizing the dark energy, and they expected it was neutralizing it perfectly, as a −1 neutralizes a +1. Such perfect cancellations were not unknown in physics. Far from it: they were a feature of the symmetries of the universe.

In the 1990s, astronomers were able to measure dark energy directly and found that the actual cosmological constant was different from the value predicted, and so was what physicists call "technically unnatural." In fact, it was much, much smaller than the predicted value—120 powers of ten less than that value. Yet it was incredibly precise. Tip the scale one one-hundredth of a decimal place in one direction and the universe would expand far too rapidly for galaxies to form. Tip it one one-hundredth of a decimal place in the other direction and the universe would collapse a fraction of a second after it appeared. If that precision had represented a perfect cancellation of the dark energy, it might not have been particularly remarkable. Instead, it turned out to be a very near but not quite perfect cancellation. This balance between the universe's expanding and contracting forces was slightly asymmetrical, and strangely, all evidence is that this slight asymmetry makes the universe possible.

Considered outside the realm of science, this tiny asymmetry might seem to be evidence of intelligent design, although, one can't help but think, by a designer who had conceived an altogether elegant table and, upon making it, found he needed to slip a twice-folded paper napkin under one leg to keep it level. It has been called the most disturbing example of cosmic fine-tuning, a "put-up job," and a big fix.[7] Considered within the realm of science, it might be a fluke, and for you and me, a very lucky one. Or, it might be evidence for a multiverse.

Specifically, this asymmetry might be evidence that most universes would be inhospitable—short-lived, courtesy of not enough dark energy, or mostly empty space, courtesy of too much—and evidence that within the subset of universes that allowed for life, conditions in the vast majority (ours included) would be met by the slimmest of margins. In other words, evidence that we are somewhere very near the outer edge of the dartboard, exactly where the odds say we should be.* In fact, in

---

* There is another explanation for the values of the constants. It also depends on odds, but odds of a different sort. In 1995, cosmologist Edward Harrison, then at the University of Massachusetts, speculated that our universe is artificial, created by an intelligence superior to ours and existing in a "mother" universe whose physical constants are similar to our own. His thinking rested on the supposition that suitably advanced civilizations, driven by a creative impulse, will wish to produce child universes and will have the means to do so. From this it follows, he wrote, that universes unfit for life cannot produce the sufficiently advanced civilizations necessary to spawn child universes. But universes fit for life *can* produce such civilizations, and these, in turn, may

1987 Steven Weinberg, using anthropic reasoning, had predicted a narrow range of values for the cosmological constant, simply by choosing values that would allow a universe with life.

LIFE IN THE MULTIVERSE

Anthropic reasoning was premised on the assumption that the values of the Standard Model and cosmological parameters were the only ones possible for a universe that allowed life. In time the assumption became a given. Many papers appeared showing that a small tweak of a value or a parameter would turn a life-allowing universe into a barren one. But in 2005, three physicists—Roni Harnik, Graham Kribs, and Gilad Perez—wondered if there might be a counterexample—that is, a value of the Standard Model or cosmological parameter that, if changed, might still yield a universe that allowed life. As to the particular value or parameter to change, the obvious candidate was the weak force.

The weak force governs the process of radioactivity. It also governs nucleosynthesis, the process that occurred in our uni-

---

produce child universes. Since universes whose constants do not support life are not reproduced, there will be relatively few of them. Since universes whose constants do support life multiply and multiply again, there will be a great number of them; and as the multiverse grows older that number will grow, with fertile universes coming to greatly outnumber sterile universes. So, since most universes are conducive to life, it follows that we are not living in a privileged place, but just another universe—a typical one. Furthermore, if at most times in the history of the multiverse there are a great number of universes, then we are not living at a privileged time either. Harrison had taken the problem presented by the physical constants and turned it upside down. And perhaps without intending to, he had also supplied a reason to consider our universe, which he hypothesized might be a product of random mutation and natural selection, as living. (Harrison, *Cosmology*) verse's first three minutes and produced the hydrogen and helium that later formed stars. Contra its name, the weak force is very strong, some 10^{32} times stronger than gravity. And like the cosmological constant, the actual, detected value is far stronger than the calculated value, and so technically unnatural. The most favored explanation for the actual value involves theoretical particles that may be discovered with the Large Hadron Collider at CERN near Geneva, Switzerland, but at present no one knows why the weak force is so strong, and many physicists—revisiting that calculated value—question why there's a weak force at all. Harnik, Kribs, and Perez turned that question into a thought experiment. They didn't imagine merely a universe with a slightly stronger or weaker weak force; they imagined a universe without one.[8]

Most physicists who made adjustments in values and force strengths made one adjustment at a time. Certainly, as mentioned earlier, a small tweak of the strength of the weak force—a tweak in either direction—would mean a lifeless universe. What Harnik, Kribs, and Perez quickly realized was that if they wanted a universe with no weak force that still allowed life, like an audiophile seeking a stereo sound with no distortion, they would have to make several adjustments at once.

THE NECESSITY OF STARS

If the physics here was so speculative as to be fringe, the biology was decidedly conservative, especially by the standards of some of the ideas described in previous chapters. It was biology that needed carbon.

Although the fundamental needs of life based in chemistry are known, the stages a universe needs to go through to produce that life are not. Nonetheless, most scientists would concede that at the very least, that universe must have stars.

Stars in our universe are made of hydrogen and helium, and both were created a few seconds after the big bang by nucleosynthesis, a process, enabled by the weak force, involving reactions that turn protons into neutrons and neutrons into protons. Without a weak force, nucleosynthesis in our universe would be impossible, hydrogen and helium could not be created, there would be nothing to make stars with, and the game would be over before it began. But Harniz, Kribs, and Perez realized that a universe with no weak force might still have stars—if certain initial conditions early in that universe's history were different from the initial conditions in our universe.

One such condition was the ratio of matter to antimatter. In our universe, the heat radiation released after the big bang created enormous amounts of matter and antimatter, and slightly more matter than antimatter. As the universe cooled, most of the antimatter and matter annihilated each other, leaving an excess of matter. If a universe with no weak force had begun its existence with a different matter–antimatter ratio, it might still manage nucleosynthesis shortly after its big bang. It would, however, be nucleosynthesis of a different sort. It would produce not the common form of hydrogen (with a nucleus made of a single proton) but the form called hydrogen 2 (with a nucleus made of a proton and a neutron). Hydrogen 2 could go to make stars, but stars different from those we know. Sun-like stars in our universe fuse hydrogen nuclei (protons) to make helium 4 nuclei (two protons and two neutrons). Stars in a universe without a weak force would fuse hydrogen 2 nuclei with a proton to produce helium 3, with a nucleus of two protons and a neutron. Such stars would be smaller and cooler than stars like our Sun, but they would nonetheless be capable of warming any planets and moons orbiting them. Moreover, they could burn for 7 billion years—long enough, if life on Earth is any guide, for organisms to arise on those planets and moons. And they would have internal processes to forge elements like oxygen, carbon, and nitrogen, the elements necessary to familiar life.

Which brings us to the remaining requirement for a life-bearing universe. The elements forged inside stars must have a means to be dispersed through space, so that they might make landfall on planets and other bodies hospitable to complex chemistry. In our universe, the elements forged within stars are dispersed when those stars explode as supernovas, and most supernovas result from collapsing stars. The explosions are caused by shock waves from the star's core, and the shock waves are sustained by neutrinos produced by the weak force. With no weak force, many massive collapsing stars would simply fizzle out, keeping their synthesized elements inside. But our universe has another kind of supernova. Its explosions, which are thermonuclear and triggered by accretion, can also disperse those elements. Such supernovas and their attendant explosions require no weak force, and they could occur in a universe without one.

Harnik, Kribs, and Perez admitted that the life in a universe without a weak force would have to make accommodations rather unlike the accommodations life has made to our universe. The traditional habitable zone of a planet orbiting the smaller, cooler stars of such a universe would be nearer those stars. In addition, the stars of a universe with no weak force could synthesize only traces of elements heavier than iron, and because they couldn't synthesize even small amounts of very heavy elements like uranium (one source of our planet's internal warmth), a planet's internal heating and plate tectonics would need to arise from a process other than radioactive decay. But as we've seen, other processes are available to heat a planet from within. As for the uranium, life probably wouldn't miss it. Biochemistry like that we know would have all the chemicals it needed to make simple organisms that would metabolize, reproduce, and evolve.

Harnik, Kribs, and Perez hypothesize that such organisms might even evolve into intelligent beings. Such beings might discover that their universe is governed by three fundamental forces, posit a multiverse, and reason that their universe is typical of the subset of universes that allow life. They would not think of themselves as fundamentally weird. (It's a good bet that no one thinks of himself or herself as fundamentally weird.) But they would think of us—living in a universe with a fourth force that is both technically unnatural and not necessary to life—as very weird indeed. And unless there's an explanation for the strength (and very existence) of our universe's weak force, they would have a point.

MORE ADJUSTMENTS, MORE UNIVERSES

A few years after Harnik, Kribs, and Perez published their work on a universe without a weak force, another team of physicists— Robert Jaffe, Alejandro Jenkins, and Itamar Kimchi—performed a similar thought experiment. By this time it had become nearly a given that any change in the masses of quarks (and hence the masses of protons and neutrons) would make for a universe hostile to life. Jaffe, Jenkins, and Kimchi wondered whether they could find an exception—that is, a change in the masses of quarks that could also yield a universe that *allowed* life.

Changing quark masses, even in theory and even for theoretical physicists, is no simple matter, and it's not surprising that the paper in which Jaffe, Jenkins, and Kimchi describe their thinking is thirty-three pages sprinkled liberally with discussions of Yukawa couplings and Higgs vacuum expectation values. Nonetheless, its conclusion was straightforward and intelligible to anyone who made it through the first week of high school chemistry. In our universe, neutrons are about 0.1 percent heavier than protons. It seemed that if protons were only slightly heavier than neutrons, you could still have hydrogen 2 and helium 3—and both could play roles in an organic chemistry that, in broad outlines at least, would be like the one used by familiar life.

Quarks come in six varieties that physicists call flavors. The "up" and "down" flavors are particularly important to the commonplace chemistry of our universe. Two up quarks and a down quark make a proton, and two down quarks and an up quark make a neutron. The four other flavors of quarks are heavier and unstable, and they quickly decay into up and down quarks. But the team found that if you could bring them into play, you could make rather more radical changes, and still have a universe that allowed life. The *strange* quark is the lightest of the heavier, unstable quarks. If you reduce its mass enough, to about the mass of the up quark, and at the same time make the down quark a good deal lighter, then you might make atomic nuclei not with protons and neutrons (the components of atomic nuclei in our universe), but with neutrons and a particle called Σ^-, or "sigma minus." A universe with atoms that had such nuclei could have stable forms of hydrogen, carbon, and oxygen, the elements necessary for organic chemistry.

All the life that developed from such chemistry might look a lot like familiar life. But that appearance would be deceiving. In a decidedly fundamental way, it would be weird. Is this then as weird as life can get? Perhaps not.

WEIRDER STILL

In 1997, Max Tegmark argued, à la Plato's realm of ideas, that mathematical structures are real. His reasoning was that on a daily basis, each of us employs a simple test of a thing's reality. We know something is real because someone else can see it too. When we apply that test to mathematical structures like geometrical theorems, they pass. Mathematical structures, said Tegmark, "satisfy a central criterion of objective experience: they are the same no matter who studies them."[9] He then carried the claim to a vastly larger scale, noting that many theoretical physicists suspect that the reason mathematics describes the universe so well is that the universe is inherently mathematical. It necessarily follows, he said, that an equation representing the theory of everything would not merely describe reality; on the most fundamental level it would be reality.

We have no such equation (at least not yet), but suppose that at some time in the future, theoretical physicists do discover one. It would immediately present us with another question: Why this particular equation, and not others?[10] Taking anthropic reasoning to another level, Tegmark offers an answer. He calls it the "ultimate-ensemble" theory. As the multiverse implies that the strengths of forces, physical constants, and dimensionality in our universe are set by chance and require no further explanation, so this theory implies that the (as yet undiscovered) equation that describes the multiverse—or as Tegmark would prefer, *is* the multiverse—was likewise set by chance and requires no further explanation. Thus, Tegmark can imagine other multiverses with other theories of everything. There might be, he suggests, a multiverse without quantum effects and a multiverse in which time is not continuous.

With only a little imagination we can add to the list. There might be a multiverse in which time flows not evenly but fitfully, a multiverse whose time flows backward, a multiverse without relativistic effects, and so on. Although Tegmark does not say so explicitly, it follows that there would be life in some of them. As with the life-allowing universes in the many-worlds multiverse and the inflationary multiverse, they would no doubt be a small subset of the whole. But they would be interesting. Their life would be weirder in a more fundamental sense even than the hypothetical life in universes without a weak force and the hypothetical life in universes whose quarks have masses different from the masses of "our" quarks.

We are rather far out on a speculative limb here, a place many theoretical physicists see no reason to venture. Some would say that Tegmark's answer to the question "why this equation?" is too easy. As Brian Greene notes, the hypothetical multiverses associated with quantum mechanics and inflationary cosmology are much more than the product of anthropic reasoning. They arose quite independent of such reasoning, from unanticipated consequences of quantum mechanics and inflationary cosmology, and so are on firmer ground. Greene has a second reservation. Theoretical physicists can hypothesize origins for the multiverses—a wave evolving via the Schrödinger equation for the many-worlds multiverse and a fluctuating inflation field for the inflationary multiverse. So far, though, no one has proposed a way that Tegmark's ultimate ensemble might have come into being.

But as long as we're on this limb, we might as well have some fun. It so happens that others have been here already, carrying anthropic reasoning still further. There are ideas of a multiverse that subsumes even the ultimate ensemble, and they proceed from the question, Why should any given universe, or for that matter any given multiverse, be fundamentally mathematical? Why not, for instance, define subsets of universes that are good or bad, or universes that are beautiful or ugly?[11] Tegmark would say that such descriptions can have no objective reality and are not scientifically meaningful. But they might nonetheless have a place within a multiverse that has been proposed, on various grounds, by philosopher Robert Nozick, philosopher David Lewis, and physicist John Barrow.[12] Each of the three has argued that our universe may be part of a multiverse that includes every possible universe.*

Recall that the traditional universe, because it contains a finite number of particle arrangements but an infinite number of particles and an infinite amount of space, must also contain all the weird life described in previous chapters, no matter how improbable, as long as it does not violate some natural law. But natural laws vary across universes. And a multiverse that contains all possible universes with all possible laws, logically, must also contain all possible life. We might reasonably expect that such a multiverse contains all the life hypothesized for the traditional universe, the many-worlds multiverse, the inflationary multiverse, and the ultimate ensemble. It must also contain a good deal more, including all the weird life of science fiction, all the animals in Robinson's *Fictitious Beasts* and all mythology and fantastic literature, as well as the vastly greater number of beings that are possible but have never been imagined.

---

* The same idea is postulated in Jorge Luis Borges's 1941 short story "The Garden of Forking Paths." A character "believed in an infinite series of times, in a dizzily growing, ever spreading network of diverging, converging and parallel times. This web of time—the strands of which approach one another, bifurcate, intersect or ignore each other through the centuries—embraces *every* possibility." (*Ficciones*, 100)

OTHER-DIMENSIONAL LIFE

In a multiverse of all possible universes and all possible natural laws, is there any kind of life that is impossible? British mathematician Gerald Whitrow suggested that the category of life that cannot exist is life in other spatial dimensions.* In 1955, Whitrow performed a set of thought experiments demonstrating the unique fitness of three dimensions for living things. It was, as several have noted, an anthropic argument. Much as Steven Weinberg's prediction constrained our universe to a specific range for the cosmological constant, Whitrow's anthropic reasoning constrained our universe to three dimensions. Whitrow noted that if space had one more dimension and gravity were unchanged, the inverse square law would be an inverse cube law, and planets would spiral into the Sun. Scaling down seemed likewise unfeasible. In a universe with two dimensions, waves could not propagate and deflect properly, and a universe with one dimension (obviously) would severely limit movement. A few years later Whitrow suggested that a universe of two dimensions would not allow the evolution of neural networks and intelligence. "In three or more dimensions," he wrote, "any number of [nerve] cells can

---

* Perhaps the best-known examples of such life were imagined by a British schoolteacher named Edwin A. Abbott. His 1888 *Flatland: A Romance of Many Dimensions* is the autobiography of "A. Square," a two-dimensional being who inhabits a two-dimensional universe that he calls, for the convenience of his three-dimensional readers, "Flatland." A. Square's education regarding worlds of various dimensions is greatly assisted through analogies with dimensions below his own—that is, Pointland and Lineland—and Abbott's readers in the third dimension are thereby invited to draw corresponding analogies to dimensions above their own. *Flatland* and most of its imaginative reworkings are mathematical fantasies and enjoyable exercises in geometry, and not intended to depict possible organisms or worlds.

be connected with [one another] in pairs without intersection of the joins, but in two dimensions the maximum number of cells for which this is possible is only four."[13]

A. K. Dewdney's *The Planiverse* (1984), which describes the physics and biology of a two-dimensional universe, goes some distance in challenging Whitrow, noting that if nerve cells are allowed to fire nerve impulses through "crossover points," they can form flat networks as complex as any in three-dimensional space. Two-dimensional minds would operate more slowly than three-dimensional ones, because in the two-dimensional networks the pulses would meet more interruptions, but they would nonetheless work. Dewdney also answers the common objection that an intestinal tract in a two-dimensional being would split it apart. He imagines a two-dimensional fish that has an external skeleton and gains nourishment by the internal circulation of food vesicles. As long as a cell is isolated, food can enter it through a membrane that can have only one opening at a time. If the cell is in contact with other cells, it can have more than one opening at a time because the surrounding cells can keep it intact.

To all such ideas you might respond that none of this matters, simply because neither we nor our descendants are likely ever to see such organisms. Whether things that we can never see or touch, and whose very existence we cannot prove, are important or unimportant is a subject with philosophical and theological implications that reach far beyond our focus here. In any case, there is a category of such life that we (or our descendants) may be able to see, touch, and measure. It is life that inhabits a specialized subset of all possible universes: universes that are simulated.

SIMULATED LIFE

In the several decades since John Conway introduced his "Game of Life," software developers have created many software programs that simulate organisms individually and collectively, and they've done so with increasing verisimilitude. Some are pure games, some are educational tools, and some, like those that ecologists use to model populations, are scientific instruments. Are simulated organisms living? If we accept the NRC report's provisional definition of life—that is, a "chemical system capable of Darwinian evolution"—the answer is no, but perhaps only because they are not a chemical system. Whether they meet the second part of the definition turns on what may be some fairly subtle distinctions. Those who claim that simulated organisms are living would say they are fully capable of Darwinian evolution; many literally are programmed for it. Moreover, they'd say, simulated organisms do all the things real organisms do: grow, consume, metabolize, reproduce, and die. In contrast, those who claim that simulated organisms are not living argue that they merely mimic these behaviors, that all the growing, consuming, and evolving—however impressive to you and me watching it happen on a screen—is virtual, not real. But still others have argued that the difference between virtual and real is increasingly indistinct. Once again, a definition of life eludes us, and it eludes us in a new way.

Suppose, however, at some point in the future simulated organisms become both living and sentient. They might regard the computer simulation as their universe, and they would understand it to be something like our own traditional universe, in which laws are the same throughout. Tegmark's conception of reality as a mathematical structure quite obviously applies here.

Computer simulations are, fundamentally, a series of mathematical manipulations that represent that computer's state. And insofar as these simulations are universes, their programs are their theories of everything.

AN UNSETTLING QUESTION

Ideas that one's existence might be nothing but the dream or imagining of another have appeared throughout literature. They are also well-trod territory in philosophy, particularly in epistemology, the field of study that asks, among other questions, how we can be sure what we know. Computer simulations, and the probability of increasingly detailed and realistic simulations, have provided a new framework for that question.

Computer power is increasing rapidly, and will likely continue to increase. If quantum computing is ever realized, processing speed will be exponentially faster. The fascination in creating simulated worlds is enticing and, it seems reasonable to expect, will continue to entice, especially as those worlds grow ever vaster in scope and richer in detail.

For the moment, then, let's allow our imaginations free reign. Members of a civilization with technology millennia in advance of ours might be able to create simulated universes, and assuming no technical obstacle, they might also be able to create inhabitants who are sentient and self-aware. The scenario of computer-generated alternate realities is common in science fiction, and it invites a question. To borrow the nearest cultural referent, how do we know we're not living in the Matrix? That is, how do we know we are not simulated beings existing in a simulated world created with technologies vastly in advance of our own, by intelligent beings—whether human, extraterrestrial, biological, or artificial? The answer, of course, is that we don't know, and it may be that we can't know. At least not easily. Of course, the simulators might deliberately reveal themselves, or they might get careless. ("Hi. Some of you may be wondering why the sky turned to static yesterday. This is kind of embarrassing, but . . .") But there might be another way. Even simulators very careful to hide their tracks might, from time to time, leave traces.

Suppose the simulators wanted to make it impossible for inhabitants of a simulated universe to ascertain with certainty that their universe was simulated. The simulated universe would not need to be infinite; it would need to extend outward from the inhabitants only as far as their instruments could probe. In other words, it would need to be only the size of an observable universe. If you're wondering what it might feel like to be a member of that suitably advanced civilization, try saying "only the size of the observable universe" a few times in a pleading what-I'd-really-like-for-my-birthday tone. More seriously, we may be getting cavalier with our assumptions. It's possible that even for the civilization we are imagining, an observable universe-sized simulation would be an ambitious undertaking. But it could be made less ambitious.

If the simulators wished to save on costs (or scrimp on birthday presents), they would realize they needn't simulate *all* the universe's parts, but only the parts its inhabitants happened to be looking at or otherwise experiencing at any given moment. This would mean that simulated matter on the molecular scale could be left unresolved except in places where an inhabitant was using an electron microscope, and stars and galaxies in deep space could also be left unresolved except where an inhabitant happened to be pointing an optical, infrared, or radio telescope. Of course, real molecules and distant stars and galaxies would in various ways affect the inhabitants' nearer environs, and the inhabitants with some knowledge of molecular theory and gravity would know how to detect and measure those effects. But the simulators could approximate them, and the inhabitants would be none the wiser.

Approximations, though, come at a cost. Over time they would grow into inconsistencies large enough to crash the program. To prevent a crash, the simulation would require occasional patching. Things would never be perfect. Even with patching, if the inhabitants looked hard enough they might detect inconsistencies, small tears in the cosmic scenery. Exactly what would those tears look like? John Barrow thinks they might look like that (possible) small change, 6 billion years ago, in the strength of the electromagnetic force[14]—a prospect that may raise the possibly simulated hairs on the back of your possibly simulated neck.

To most of us, the idea that we are simulations is fairly abhorrent, but that is not a reason to think it unreal. In fact, the Oxford philosopher Nick Bostrom makes a persuasive case. Bostrom reasons that members of a society with a sufficiently advanced technology, a society that can create simulated universes, will create a great many. Sooner or later inhabitants of simulated universes will outnumber inhabitants of actual universes, and their population will only increase over time. Enter, once again, the principle of mediocrity. You can probably see where this is heading. If we do not know whether our universe is real or simulated but we have reason to believe that simulated universes vastly outnumber real ones (suppose, for instance, there are a thousand simulated universes for every real one), then we've no choice but to conclude that odds are we are living—er, "living"—in a simulated universe.

We can't know the sorts of universes or inhabitants that society with the suitably advanced technology prefers to create, but we might with some reason expect that its members wish to be edified and entertained. It's reasonable to assume that if they can begin simulations, they can also end them—and they may do so for many reasons, including waning interest. So, if that possible change in the electromagnetic force 6 billion years ago has you worried, you might take it as incentive to keep the simulators interested. Exactly what might the simulators find interesting? Obviously we can't know, but if they are at all like the players of computer games I've seen, they bore easily, they seldom reward good behavior, and they almost never reward passivity. If we'd like our universe to continue running, perhaps we all should start taking more risks.

Readers who haven't put down the book to make a reservation to bungee jump or take up extreme downhill skiing will be comforted to know that simulators might program the simulation for risk in many ways, most of which pose no threat to our well-being. One of these brings us back to weird life. Let's back up on the speculative branch a bit, and assume that we inhabit a real universe. But let's also assume that as a more or less inevitable by-product of technological advances, computer simulations will continue to become ever more realistic. Simulations invite risk precisely because there is nothing real *at* risk. All simulators are inclined to experiment, to explore extremes, to push boundaries. This is true for a game player breaking the world land speed record on a twisting stretch of autobahn. It is also true for an entomologist manipulating virtual populations of monarch butterflies.

Might the same inclination toward risk be manifest in the organisms themselves? Perhaps. It may not be too much to expect that simulators interested in biology will have a preference for subroutines that generate simulated weird life. Even now, biologists simulate organisms and their parts; the processes of protein folding and binding, for instance, are routinely simulated with parallel and distributed computing.[15] Simulated organisms in computer games, a realm mostly removed from scientific inquiry, are commonplace, and a quick census would likely show that the weird (meaning dragons and the like) would greatly outnumber the familiar.*

However, merely observing simulations and making an adjustment now and then is likely to prove a dull pastime. Sooner or later, simulators would wish to engage their simulations directly. This wish, in a manner that no doubt will one day seem primitive, is being answered even now. Will Wright, the designer of the video game simulations *SimCity*, *The Sims*, and *Spore*, is designing a game he calls *HiveMind*. It is a set of cross-platform, online applications designed to turn a gamer's everyday life into an interactive experience by tapping into personal information on phones, tablets, social networks, and computers. It will offer an experience, so Wright claims, that will merge the virtual with the real.[16]

Whether this is the first iteration of such a merging or a false start, it seems clear that the boundary between the real and the virtual is likely to grow ever more porous. Assuming no technical impediments, at some time in the future we'll see two-way traffic between real and simulated environments. From the previous we might draw an interesting conclusion. If the impulse to simulate weird life is significant, and if computing power continues

---

* If we return for a moment to Bostrom's argument for the probability that we are simulations, and if those simulators share a weakness for the weird, then it follows that we are weird life—weird, that is, in the view of our simulators.

to increase to the degree that real persons can visit simulations and simulations can visit the real, then it follows that we or our descendants will encounter weird life directly—a prospect that yields to imaginings of griffins leaving beak marks on our ankles and unicorns following us home.

A BIT OF PERSPECTIVE

We've traveled a long way from speculation about bacteria with arsenic in their DNA and desert varnish as living. It may be time to catch our breaths and consider the lines of thought that brought us here. The attentive reader may have noticed that ideas for the weirdest sorts of weird life did not originate with biologists or even, for that matter, with astrobiologists. They came from scholars and practitioners in other fields. The hypotheses of life in other universes were formulated by theoretical physicists (Harnik, Kribs, and Perez; and Jaffe, Jenkins, and Kimchi). Ideas of life in the vicinity of black holes and the atmospheres of white dwarf stars were conceived by astrophysicists (Adams and Laughlin). Hypotheses of life surviving through eternity were developed by a mathematician and theoretical physicist (Dyson), who also supplied us with what may be the broadest definition of life so far. Of the many ideas of weird organisms from science fiction, two that are notably well grounded in science are from a physicist turned aerospace engineer (Forward) and a professional astronomer (Hoyle). Even the relatively conservative hypothesis of hydrogen-breathing dirigibles was proposed by a physicist (Saltpeter) and a planetary scientist turned astrobiologist (Sagan).

Many biologists who have considered alternative forms of life are inclined to agree with Norman Pace, who, you'll recall, suspects that if life exists elsewhere in the universe, it has a biochemistry much like that of life we know. Certain astro-biologists—Schulze-Makuch and McKay, for instance—do have hypotheses for organisms that use another biochemistry; but by comparison with ideas of life on crusts of neutron stars and of life in the vicinity of black holes, they seem fairly tame. Clearly, there are two very different levels of speculation here, and between them a wide gap that needs explaining.

It's reasonable to suppose that the reason for some of the gap is the nature of the fields of study (physics and astronomy on the one hand, and biology on the other), the perspectives that come with study in those fields, and the sort of intellect those fields attract to begin with. On the one hand, and to generalize, theoretical physicists are likely to be less interested in the particulars of a given phenomenon than they are in the underlying principles those particulars represent. Astronomers are necessarily cognizant of billions of stars as potential suns, the billions of planets now estimated to be orbiting them, and the enormous timescales over which the universe has existed and is likely to continue to exist. To a biologist pessimistic about the likelihood of weird life, a theoretical physicist might say that the particulars of biochemistry and chemistry are not as important as life's most basic needs: energy and matter. And an astronomer would insist that those needs are met in many places in our universe (and if they exist, others), and have been met for billions of years.

On the other hand—and to generalize further—a biologist is more likely than is a theoretical physicist to be aware of the fantastically complex chemical reactions that occur within a living cell, and the improbably long series of steps, many still not fully understood, that led from amino acids to that cell. To a physicist or astronomer optimistic about the probability or inevitability of weird life, a biologist might counter that he fails to appreciate the intricacies of chemistry and biochemistry that would be necessary to produce it.[*]

The gap is a reminder that scientists have taken the concept of weird life seriously only recently. If it may be said to represent a new field of study, it is a field that, as was once said of astrobiology, has no subject. Or no subject *yet*.

---

[*] There is a difference of opinion split along the same disciplinary lines on the question of the probability of extraterrestrial intelligence. It was articulated perhaps most succinctly in a debate between Carl Sagan and Ernst Mayr, with Sagan taking the view that extraterrestrial intelligence was common, Mayr that it was rare and perhaps unique to our own species. Mayr observed that Sagan and others who estimated large numbers of extraterrestrial civilizations used a flawed reasoning attributable to their professional backgrounds: "When one looks at their qualifications, one finds that they are almost exclusively astronomers, physicists and engineers. They are simply unaware of the fact that the success of any SETI effort is not a matter of physical laws and engineering capabilities but essentially a matter of biological and sociological factors. These, quite obviously, have been entirely left out of the calculations of the possible success of any SETI project." (http://www.astro.umass.edu/~mhanner/Lecture_Notes/Sagan-Mayr.pdf)

Epilogue

At present, no one has discovered an example of weird life, and it's possible that no one ever will. For these reasons the subject resists the kind of tidy conclusion customary in an epilogue. But I can at least address the prospects for searches.

Although Davies and others have advocated a program for a set of dedicated and systematic searches for weird life on Earth, at present there is no such program. As described in the text, individual studies have sought weird life on Earth deliberately, and one may yet find it; but it's also possible that the scientists who discover weird life on Earth do it, as it were, accidentally, while looking for something else. As to *in situ* searches for weird life on other worlds, they are a long way off at best. The only mission now under way to perform on-site study of any planet or moon's surface is NASA's Mars Science Laboratory, and it is designed to detect not life itself, but merely the conditions that would make life possible. Moreover, because mission scientists are defining those conditions as the ones suitable to life we know, the mission could fail to notice evidence of weird life. Although

ESA has plans for a mission that will visit Jupiter and its moons, it is not designed specifically to seek life. No missions to the surfaces of two worlds high on the wish lists of astrobiologists—namely, Saturn's moons Enceladus and Titan—are currently being developed.

Norman Pace observes that the first evidence for extraterrestrial life is likely to come from spectroscopic detection of chemical disequilibrium in the atmospheres of planets or moons in our Solar System or, more probably, outside it.[1] Here, the numbers of both the research programs and the subjects being studied may vastly increase the odds. The mission of the Kepler space observatory, originally scheduled to end in November 2012, may be extended by two years, and perhaps two years beyond that. The mission of the COROT spacecraft (overseen by France's National Center for Space Studies in cooperation with ESA), like the mission of Kepler, is the detection of planets by the transits, and it is ongoing. Roughly eighty programs for detecting extrasolar planets with Earthbound telescopes are either ongoing or in development,[2] and many of them can perform spectroscopic analysis of planetary atmospheres. In advocating for a search for weird life, the NRC report noted that although the biosignatures of weird life in planetary atmospheres would be different from the biosignatures of life like our own, they could be detected just as readily.[3] Of course, as mentioned in the text, without on-site study or a sample return it would be impossible to know for certain whether they were real biosignatures, or merely the product of some very unusual chemistry.

In the meantime, as we've seen, scientists are speculating, hypothesizing, and building models of weird biochemistry. Such thinking is valuable simply as an exercise, as noted some 300 years ago by Christiaan Huygens:

If anyone shall gravely tell me that I have spent my time idly in a vain and fruitless inquiry after what I can never become sure of, the answer is that at this rate he would put down all natural philosophy, as far as it concerns itself in searching into the nature of things. In such noble and sublime studies as these, 'tis a glory to arrive at probability, and the search itself rewards the pains.[4]

But thinking seriously about weird life has other uses too, as it helps scientists escape and overturn Kuhnian paradigms, expect the unexpected, and prepare themselves to recognize what their training and experience have not taught them to recognize.

It is possible that SETI will detect that long-sought signal. Otherwise, the first extraterrestrial life that humans encounter and can prove to be living is likely to be found with sophisticated remote-controlled spacecraft performing on-site studies, or sample-return missions. Since simple single-celled life arises more easily than complex life, odds are that the life encountered will be microbial. Suppose that it is weird. Upon hearing the news and seeing an image of a microbe that seems indistinguishable from any other, some will surely wonder what the fuss is about. To fully appreciate the discovery's importance, many of us will need a refresher in biology—most probably a good thing.

Already, hypothesizing about weird life has contributed, albeit indirectly, to our appreciation and understanding of the part of the natural world we know exists. Many scientists who have hypothesized weird organisms made a case for their feasibility by noting parallels with familiar life. The layering of desert varnish (that candidate for weird life) is like the layering of stromatolites. Sagan and Saltpeter's hypothesized Jovian ecology of hydrogen-breathing dirigibles is borrowed from the known ecology of microfauna in sunlit waters of Earth. And Dyson and Hoyle's ideas of living interstellar clouds of dust grains and complex molecules organized by electromagnetic forces are modeled on neurotransmitters in animal brains. To make cases for the viability of these parallels, they have had to revisit the familiar.

The features of life we know continue to surprise biologists on a regular basis. And familiar life in all its variety still inspires imagination, reminding us how much of it remains undiscovered and unstudied. During an age in which many are increasingly disconnected from the natural world, an appreciation of that world from a fresh perspective is no small matter. If that is all the search for weird life ever gives us, it may be enough.

Glossary

aerosol A fine aerial suspension of liquid (mist or fog) or solid (dust or smoke) particles.

amino acid An organic compound; an essential component of the protein molecule.

Archaea One of the three largest groups, termed "domains," in the three-domain system, the taxonomy introduced in the 1960s by microbiologist Carl Woese. The majority of extremophiles, as well as the most extreme extremophiles, belong to this domain. In the five-kingdom system, the same organisms are classified within the kingdom Bacteria as "archaebacteria."

astrobiology Also termed *exobiology* and, less commonly, *bioastronomy*. The study of extraterrestrial life (all of which, as this book goes to press, is hypothetical) and the conditions that might make such life possible. More recently, the study of life in the context of the universe.

bacteria Microscopic single-celled organisms lacking nuclear membranes around the genes and/or nucleus. All bacteria are prokaryotes. In Carl Woese's three-domain system, they constitute the domain Bacteria.

bioastronomy See *astrobiology.*

biochemistry The chemistry of biological substances and processes.

biology The study of living organisms and life processes, including their structure, functioning, growth, origin, evolution, and distribution.

biosolvent A liquid medium that allows and facilitates chemical reactions conducive to life.

biosphere All life on Earth, existing in a region measured from upper levels of the atmosphere to several kilometers below the planet's surface. The precise upper and lower limits of the biosphere have yet to be determined.

catalyst A substance that speeds the rate of a reaction by providing a lower-energy alternative pathway.

cell The smallest structural unit of an organism that is capable of independent functioning, consisting of one or more nuclei, cytoplasm, and various organelles, all surrounded by a semipermeable membrane.

chemistry The study of the composition, structure, properties, and reactions of matter.

chirality Handedness; a configuration of any molecule that prevents it from interacting biochemically with mirror images of itself.

chromosome A (usually rod-shaped) structure that carries genes.

convergent evolution An increase in the degree of similarity between two or more unrelated species as they evolve.

cosmological constant A term, introduced by Einstein into the field equations of general relativity, that allowed for a static universe, neither expanding nor contracting.

cytoplasm The contents outside of the nucleus and enclosed within the cell membrane of a cell. The cytoplasm is clear in color and has a gel-like appearance. Composed mainly of water, it also contains enzymes, salts, and various organic molecules.

DNA Deoxyribonucleic acid. The fundamental hereditary material of all living organisms that biologists know of; the large molecule that composes genes.

ecology The study of the interaction of organisms with their environment.

ecosystem The organisms living in a particular environment, and the physical part of the environment that affects them and/or impinges on them.

Eukarya One of the three largest groups, termed "domains," in the three-domain system, the taxonomy introduced in the 1960s by microbiologist Carl Woese.

eukaryotes Organisms (both microscopic and macroscopic) made of cells with membranes enclosing their genes and/or nucleus. They include protists (such as algae and slime molds) as well as fungi, plants, and animals. Compare to *prokaryotes*.

exobiology See *astrobiology*.

extremophile An organism that thrives under extreme environmental conditions of heat, pressure, pH, and so on. They include thermophiles, hyperthermophiles, psychrophiles or cryophiles, barophiles or piezophiles, acidophiles, alkaliphiles, halophiles, and radiophiles.

Gaia hypothesis The proposition that all organisms and their inorganic surroundings on Earth compose a single and self-regulating system that maintains conditions suitable for life.

gene The basic unit of heredity.

Goldilocks zone See *habitable zone*.

habitable zone Also termed *Goldilocks zone*. The region around a star within which it is theoretically possible for a planet with sufficient atmospheric pressure to maintain liquid water on its surface. More recently, the region anywhere (including interiors of planets and moons and other celestial bodies) where water might exist in liquid form.

hypothesis A scientific proposition that is supported by observational evidence and purports to explain a given phenomenon or set of phenomena. A hypothesis is neither as comprehensive nor as well established as a *theory*, although a set of related hypotheses may, over time, come to constitute a theory.

last universal common ancestor (LUCA) The theoretical organism from which all life is descended.

mesophile An organism that grows best in moderate temperatures, typically between 20°C and 45°C.

metabolism The sum of the physical and chemical processes that occur in a living organism.

methanogen An organism capable of producing methane from the decomposition of organic material.

microbe A microorganism.

microbial community A highly organized and well-integrated system of microbes that modifies its environment chemically and physically.

microbiology The scientific study of microscopic organisms.

multiverse The (hypothetical) set of all universes that results from one of several scenarios, including the "many-worlds" approach to quantum mechanics; versions of the "inflationary" hypothesis, in which universes are sprung from separate big bangs into different regions of space-time; and the "ultimate-ensemble" hypothesis, in which our universe is a mathematical structure and other universes are other mathematical structures.

organelle An organized structure within a cell.

organic Denoting or relating to chemical compounds containing carbon.

panspermia The theory that life on Earth and/or other suitable habitats originated on another world and arrived from outer space.

prebiotic Occurring before life appeared.

prokaryotes Organisms lacking nuclear membranes around the genes and/or nucleus. Most prokaryotes are bacteria. Compare to *eukaryotes*.

protein Any group of complex organic compounds consisting essentially of amino acids.

quantum mechanics Also termed *quantum physics*. The laws of physics that explain the behavior of the universe on very small scales (the scales of molecules, atoms, and electrons) and underlie the universe on larger scales. The laws that account in some way for vacuum fluctuations, the wave-particle duality, and various phenomena described by the Heisenberg uncertainty principle.

ribosome An assemblage of RNA and protein found in the cytoplasm of living cells and active in the synthesis of proteins.

RNA Ribonucleic acid.

shadow biosphere A hypothetical biosphere composed of weird life.

silane Any of a group of silicon hydrides, analogous to paraffin hydrocarbons and having the general formula SiH.

spore A microorganism in a dormant or resting state.

symbiosis The interaction of two organisms, typically to their mutual advantage.

synthetic biology Engineering of biological components or systems that are not known to nature; also, the reengineering of existing biological components.

taxonomy The science of classifying organisms.

theory A set of *hypotheses* that are supported by observational evidence and purport to explain a phenomenon or set of phenomena.

vertebrate Any animal with a backbone of bony segments enclosing the central nerve cord. The five major vertebrate groups are fishes, amphibians, reptiles, birds, and mammals.

vesicle A bladderlike cavity or sac, especially one filled with fluid.

virus An organized set of chemicals that is capable of reproduction and evolution but not of metabolism, and so is considered by most biologists not to be a living organism.

Notes

PROLOGUE

1. Margaret W. Robinson, *Fictitious Beasts: A Bibliography* (London: Library Association, 1961).
2. Wilson, *Future of Life*, 14.
3. Bryson, *Short History*, 322.
4. Lee, *Vegetable Lamb of Tartary*.
5. Hooke, *Micrographia*, 210.
6. Sattler, Puxbaum, and Psenner, "Bacterial Growth."
7. Funch and Kristensen, "Cycliophora Is a New Phylum."
8. National Research Council. Committee on the Limits of Organic Life in Planetary Systems, *The Limits of Organic Life in Planetary Systems* (Washington, DC: National Academies Press, 2007).

CHAPTER ONE: EXTREMOPHILES

1. Interview with the author, March 19, 2010.
2. In truth, my joshing here is unearned. There is a great deal to be excited about, especially recently. Samples of sediment from off the coast of Newfoundland suggest that bacteria and archaea may survive 1,600 meters below the seafloor, thus extending the

biosphere deeper than many had imagined possible. Further, such sediment may contain as much as two-thirds of the Earth's prokaryotic biomass by weight. See Roussel et al., "Extending the Sub-Sea-Floor Biosphere."

3. Interview with the author, March 19, 2010.

4. Edmond and Von Damm, "Hot Springs," 86.

5. Bryson, *Short History*, 274.

6. Piccard and Dietz, *Seven Miles Down*, 173.

7. Gross, *Life on the Edge*, 23.

8. The oxygen, it should be noted, is produced by phytoplankton, meaning that a part of this process depends, indirectly but ultimately, on photosynthesis.

9. Brock, "Road to Yellowstone."

10. Stanier, Dondoroff, and Adelberg, *Microbial World*.

11. Kuhn, *Structure of Scientific Revolutions*, 63–64.

12. Ibid., 64.

13. MacElroy, "Some Comments."

14. Stetter, "Hyperthermophilic Procaryotes."

15. If one defines extremophiles as organisms adapted to environments hostile to life and takes the long view, you and I might well be included. The first life on Earth was anaerobic; it would have found oxygen to be toxic. In fact, although oxygen makes possible a more efficient metabolism, that efficiency may come at a cost. Cellular damage from oxygen has been implicated as a contributing cause of aging and cancer.

16. "Although they are rare, some environments with liquid water do not appear to support life; they include water over 400°C at submarine hydrothermal vents [and] high-brine liquid water found in sea-ice inclusions at −30°C." But even in these extremes, microbes have been known to survive. (National Research Council, *Limits of Organic Life*, 31)

17. Kidd, *Adaptation of External Nature*.

18. Chung, "How Bug Extends."

19. National Research Council, *Limits of Organic Life*, 35.

20. There is at least one interesting exception (Wharton and Ferns, "Survival of Intracellular Freezing").

21. National Research Council, *Limits of Organic Life*, 35.

22. It should be noted that most of the salt in the water of the oceans is sodium chloride; the saline chemistry of the Dead Sea is more complex, also including significant proportions of magnesium chloride, calcium chloride, and potassium chloride.

23. In his retirement, Volcani resumed his studies of Dead Sea biota. In the late 1990s, he opened bottles of enrichment cultures that he had collected half a century earlier and found live microorganisms, at least one of which had never been identified. (Ventosa, Arahal, and Volcani, "Microbiota of the Dead Sea")

24. Interview with the author, March 19, 2010.

25. Marine Biological Laboratory, "Life at the Extremes." The Phoenicians are said to have called the Rio Tinto the "River of Fire." NASA, somewhat more prosaically, calls it a "Mars Analog," and is conducting research (the Mars Analog Research and Technology Experiment) that seeks microorganisms, akin to those that might live beneath the Martian surface, in the rock and ore several hundred meters beneath the river.

26. Fountain, "Date Palm Seed."

27. Cano and Borucki, "Revival and Identification of Spores."

28. Vreeland, Rosenzweig, and Powers, "Halotolerant Bacterium."

29. Kelvin, "On the Origin of Life," 202.

30. Pedersen, "Deep Intraterrestrial Microbial Life."

31. Belozerskaya et al., "Extremophilic Fungi from Chernobyl."

32. Hart, "Hydrothermal Vents."

33. Rothschild and Mancinelli, "Life in Extreme Environments."

34. These include—and some technical language is necessary here—the tricarboxylic acid cycle, glycolysis, and synthetic pathways for the construction of amino acids and sugars.

35. We should not be surprised to learn that Darwin put forth much the same hypothesis, writing, "I should infer from analogy that probably all the organic beings which have ever lived on this earth

have descended from some one primordial form, into which life was first breathed." (Darwin, *Origin of Species*, 380)

36. In fact, it is a reasonable modification of a much-used term. All individuals of a given species, for instance, share a "last common ancestor" that is also the first member of that species.

CHAPTER TWO: A SHADOW BIOSPHERE

1. Darwin, *Life and Letters*, 498.

2. Paul C. W. Davies, Steven A. Benner, Carol E. Cleland, Charles H. Lineweaver, Christopher P. McKay, and Felisa Wolfe-Simon, "Signatures of a Shadow Biosphere," *Astrobiology* 9, no. 2 (2009): 241–49.

3. Maher and Stevenson, "Impact Frustration."

4. Slater, "Biological Problems."

5. Erwin, "Tropical Forests," 74.

6. Wilson, *Future of Life*, 14. The authors of a 2011 study, using patterns within the taxonomic classification system, estimated the total number of species on Earth to be 8.7 million, give or take 1.3 million. (Mora et al., "How Many Species")

7. Quoted in Wilson, *Diversity of Life*, 142.

8. Pace, "Molecular View of Microbial Diversity."

9. National Research Council, *Limits of Organic Life*, 29.

10. Wilson, *Future of Life*, 20.

11. "Life in the Universe," 3.

12. Navarro-González et al., "Mars-Like Soils."

13. Stevens and McKinley, "Lithoautotrophic Microbial Ecosystems"; Chapelle et al., "Hydrogen-Based Subsurface Community"; and Lin et al., "Planktonic Microbial Communities."

14. Sogin, "In Search of Diversity."

15. Kaufman, "Second Genesis on Earth."

16. "[Our analyses] demonstrated that intracellular AsO_4^{3-} was incorporated into key biomolecules, specifically DNA." (Wolfe-Simon et al., "Bacterium That Can Grow," 3)

17. Zimmer, "This Paper Should Not."

18. Ibid.
19. Reaves, "Absence of Arsenate."
20. Pikuta et al., "Bacterial Utilization."
21. "Size Limits of Very Small Microorganisms."
22. Folk, "SEM Imaging."
23. Uwins, Webb, and Taylor, "Novel Nano-organisms."
24. Kajander and Ciftcioglu, "Nanobacteria."
25. Schieber and Arnott, "Nannobacteria as a By-Product."
26. Young and Martel, "Truth about Nanobacteria."
27. Asaravala, "Are Nanobacteria Making Us Ill?"
28. Ibid.
29. Smith, "Nanobes."
30. Darwin, *Voyage of the Beagle*, 13.

CHAPTER THREE: DEFINING LIFE

1. Interestingly, it may be that Earth orbits a bit outside a sweet spot within the habitable zone, and that life in general would prefer a warmer planet. Biodiversity increases in warmer and wetter climes, and it is greatest in equatorial rain forests, the warmest and wettest places on Earth's surface. There is no obvious reason to think the trend wouldn't continue into places steamier still, if there were any. (Impey, "New Habitable Zones," 24)

2. Schenk et al., "Ages and Interiors."

3. Impey, "New Habitable Zones," 25.

4. Hussmann, Sohl, and Spohn, "Subsurface Oceans," 258–73.

5. Impey, *Living Cosmos*, 205.

6. Huxley, "On the Physical Basis of Life," 130–65. In the magnificent conclusion of *Origin*, Darwin allowed for a vital force, writing, "There is grandeur in this view of life, with its several powers, having been originally breathed into a few forms or into one" (*Origin of Species*, 384). In the second edition, the phrase "having been originally breathed" is followed with "by the Creator." Darwin seems to have meant the original wording only figuratively, and he lamented his later revision as disingenuous. "I have long

regretted that I truckled to public opinion, and used the Penta-
teuchal term of creation, by which I really meant 'appeared' by
some wholly unknown process. It is mere rubbish, thinking at
present of the origin of life; one might as well think of the origin
of matter" (letter to J. D. Hooker on March 29, 1863, as reproduced in Darwin, *Life and Letters*, 498).

7. Pirie, "Meaninglessness," 11–22.

8. Keosian, *Origin of Life*, 16.

9. Horowitz, "Biological Significance," 3.

10. For a summary of the view that viruses are nonliving, see Moreira
and López-García, "Ten Reasons to Exclude Viruses." For an argument that definitions of viruses as nonliving amount to "dogma,"
see Bandea, "Origin and Evolution of Viruses."

11. Dawkins, *Selfish Gene*, 191.

12. Interestingly, Dawkins allows that memes can be produced in
nonhuman cultures as well. Songbirds' songs are mimicked imperfectly, with the imperfections copied by other birds.

13. Gardner, "Fantastic Combinations."

14. Davies, *Eerie Silence*, 81.

15. In his book *Origins of Life*, Freeman Dyson did just that.

16. National Research Council, *Limits of Organic Life*, 7.

17. Ibid., 6.

18. In 1995, biologist Lynn Margulis and science writer Dorion Sagan
recycled Schrödinger's title *What Is Life?*, reframing the question
with recent knowledge.

19. Cleland, "Life without Definitions."

20. Pittendrigh, Vishniac, and Pearman, *Biology and the Exploration
of Mars*, 5.

21. Gribbin and Gribbin, *James Lovelock*, 139–40.

22. In 1965, Earth-based spectrometry analyzed Mars's atmosphere
and found it to be mostly carbon dioxide, indicating a state of
thermodynamic equilibrium. Although a later finding would
overturn this result (in 2004, the *Mars Express* mission discovered traces of atmospheric methane that may or may not have

a geochemical origin), at the time Lovelock regarded the spectrometry results as persuasive evidence that Mars was lifeless and *Viking*'s biology experiments unnecessary.

23. Dick, *Biological Universe*, 147.

24. Cooper, *Search for Life on Mars*, 149.

25. Ibid., 126–27.

26. Ibid., 37–38.

27. Ibid., 73–74.

28. Ibid., 121.

29. Ibid., 120–21.

30. Dick, *Biological Universe*, 153.

31. Cooper, *Search for Life on Mars*, 108.

32. Dick, *Biological Universe*, 155.

33. Ibid., 158.

CHAPTER FOUR: STARTING FROM SCRATCH

1. National Research Council, *Limits of Organic Life*, 34.

2. Fox, "Life—but Not," 35.

3. Nuland, *How We Live*, 157.

4. De Duve, *Guided Tour*, 293.

5. Biophysicist Harold Morowitz writes, "To be an entity, distinguished from the environment, requires a barrier to free diffusion. The necessity of thermodynamically isolating a subsystem is an irreducible condition of life." (*Beginnings of Cellular Life*, 8)

6. Pace, "Universal Nature of Chemistry."

7. Wells, *Early Writings*, 146.

8. Cooper, *Search for Life on Mars*, 84.

9. Angier, *Canon*, 124.

10. Fox, "Life—but Not," 37.

11. Bains, "Many Chemistries," 154.

12. See, for instance, Muller, Zilche, and Auner, "Recent Advances."

13. Bains, "Many Chemistries," 154–55.

14. Ibid., 154.

CHAPTER FIVE: A BESTIARY OF WEIRD LIFE

1. Clark, "ESA Chooses Jupiter."
2. Spohn and Schubert, "Oceans in the Icy Galilean Satellites."
3. Bains, "Many Chemistries," 149.
4. Mullen, "Swimming in a Salty Sea."
5. Reyes-Ruiz et al., "Dynamics of Escaping Earth Ejecta."
6. Gugliotta, "Fountains of Optimism," 2.
7. National Research Council, *Limits of Organic Life*, 74.
8. Benner, Ricardo, and Carrigan, "Common Chemical Model."
9. McKay and Smith, "Possibilities for Methanogenic Life"; Schulze-Makuch and Grinspoon, "Biologically Enhanced Energy."
10. McKay and Smith, "Possibilities for Methanogenic Life."
11. Strobel, "Molecular Hydrogen in Titan's Atmosphere."
12. Shiga, "Hints of Life," 1.
13. Coustenis et al., "Joint NASA-ESA."
14. Shiga, "NASA Floats Titan Boat Concept."
15. Lunine, "Titan as an Analog."
16. Carter, "Anthropic Principle."
17. Bains, "Many Chemistries," 161.
18. Amato et al., "Microorganisms Isolated."
19. Imshenetsky, Lysenko, and Kazakov, "Upper Boundary of the Biosphere."
20. National Research Council, *Limits of Organic Life*, 73.
21. Schulze-Makuch and Irwin, "Reassessing the Possibility."
22. Morowitz and Sagan, "Life in the Clouds of Venus?"
23. Sagan and Saltpeter, "Particles, Environments, and Possible Ecologies."
24. Sagan, *Cosmos*, 30.

CHAPTER SIX: LIFE FROM COMETS, LIFE ON STARS, AND LIFE IN THE VERY FAR FUTURE

1. Interview with the author, July 21, 2010.
2. Ibid.

3. Allamandola and Hudgins, "Interstellar Polycyclic Aromatic Hydrocarbons," 44.

4. In summer of 2007, an international team of researchers found that particles of dust in space might form helical structures that could actually reproduce and evolve. (Tsytovich, "Plasma Crystals")

5. Dyson, "Time without End," 453.

6. Ibid., 449.

7. Ibid.

8. Quoted in Dick, *Biological Universe*, 21.

9. Maude, "Life in the Sun"; Feinburg and Shapiro, *Life beyond Earth*.

CHAPTER SEVEN: INTELLIGENT WEIRD LIFE

1. Cocconi and Morrison, "Searching for Interstellar Communications."

2. Dick, *Biological Universe*, 454.

3. "Life in the Universe," 63–64.

4. Lunine, "Saturn's Titan," 16.

5. Dick, *Biological Universe*, 434.

6. Hart, "Absence of Extraterrestrials"; Viewing, "Directly Interacting."

7. Dick, *Life on Other Worlds*, 218.

8. In 1979 the "extraterrestrial question" was the centerpiece of a conference held at the University of Maryland in College Park. Proceedings were collected in Hart and Zimmerman, *Extraterrestrials—Where Are They?*

9. Morrison, "Twenty-Five Years of the Search," 18.

10. Shostak, *Confessions*.

11. Vinge, "Coming Technological Singularity."

12. Clarke, *2001: A Space Odyssey*, 185–86.

13. West, *H. G. Wells*, 233.

14. SETI League, "Declaration of Principles."

15. Tegmark, "Multiverse Hierarchy," 8.

16. Freudenthal, *Lincos*.

17. Nagel explains as follows: "The fact that we cannot expect ever to accommodate in our language a detailed description of Martian or bat phenomenology should not lead us to dismiss as meaningless the claim that bats and Martians have experiences fully comparable in richness of detail to our own. It would be fine if someone were to develop concepts and a theory that enabled us to think about those things; but such an understanding may be permanently denied to us by the limits of our nature." ("What Is It Like to Be a Bat?" 440)

18. Hempel and Shepard, *Unleashed*.

19. Wilson, *Anthill*.

20. Drake and Sobel, *Is Anyone Out There?*, 47.

21. Dudzinski and Frohoff, *Dolphin Mysteries*, 119.

22. Gopnik, "Plant TV."

23. Ibid.

24. Dick, *Biological Universe*, 266.

CHAPTER EIGHT: WEIRD LIFE IN SCIENCE FICTION

1. Wayne Douglas Barlowe, Ian Summers, and Beth Meacham, *Barlowe's Guide to Extraterrestrials: Great Aliens from Science Fiction Literature*, 2nd ed. (New York: Workman, 1987).

2. Adams, *Hitchhiker's Guide*, 40.

3. One of the earliest examples comes from a 1930 work by British author Olaf Stapledon, in which a character explains, "In a few of the younger stars there is life, and even intelligence. How it persists in an incandescent environment we know not, whether it is perhaps the life of the star as a whole, as a single organism, or the life of many flame-like inhabitants of the star." (Quoted in Dick, *Biological Universe*, 247)

4. Forward, "When You Live upon a Star."

5. Dyson, "Time without End."

6. Hoyle, *Black Cloud*, 170.

7. Ibid., 149.

8. Dick, *Biological Universe*, 241.

9. Vernier, "SF of J. H. Rosny."

10. Darwin, *Origin of Species*, 248.

CHAPTER NINE: WEIRD LIFE IN THE MULTIVERSE

1. Bruno, *L'infinito universo e mondi.*

2. Greene, *Hidden Reality*, 311.

3. Folger, "Physics' Best Kept Secret," 7.

4. Rothman, "'What You See Is," 91. The exactness of the constants provoked intellectual crises for several scientists, including (ironically enough) Fred Hoyle. Hoyle said that his atheism was "shaken" by the discovery that if the carbon resonance level were only 4 percent lower, carbon atoms would not form. In 1981 he told an audience at Caltech, "A commonsense interpretation of the facts suggests that a superintellect has monkeyed with physics, as well as with chemistry and biology, and that there are no blind forces worth speaking about in nature." (Hoyle, "Universe," 12)

5. Bucher and Spergel, "Inflation in a Low-Density Universe."

6. Murphy et al., "Possible Evidence."

7. Davies, *Goldilocks Enigma*, 139, 149.

8. Jenkins and Perez, "Looking for Life."

9. Tegmark, "Parallel Universes," 49.

10. Tegmark, "Multiverse Hierarchy," 10.

11. Davies, *Goldilocks Enigma*, 212.

12. Nozick, *Philosophical Explanations*; Lewis, *On the Plurality of Worlds*; Barrow, *Pi in the Sky.*

13. Whitrow, *Structure and Evolution*, 200.

14. Barrow, *Constants of Nature*, 251–74.

15. Scheraga, Khalili, and Liwo, "Protein-Folding Dynamics."

16. Gaudiosi, "'Sims' Designer."

EPILOGUE

1. Pace, "Universal Nature," 805.

2. "Extrasolar Planets Global Searches (Ongoing Programmes and

Future Projects)," *The Extrasolar Planets Encyclopedia*, last updated March 13, 2012, http://www.exoplanet.eu/searches.php.

3. "Nonterran life may change the gross characteristics of planetary environments in ways that differ from influences stemming from terran life, and these differences (for example, the relative abundances of atmospheric species) may ultimately be observable over interstellar distances with astronomical facilities now on the drawing board." (National Research Council, *Limits of Organic Life*, x)

4. Huygens, *Celestial Worlds Discovered*, 10.

Works Cited

Adams, Douglas. *The Hitchhiker's Guide to the Galaxy*. New York: Harmony Books, 1979.

Adams, Fred, and Greg Laughlin. *The Five Ages of the Universe: Inside the Physics of Eternity*. New York: Free Press, 1999.

Allamandola, L. J., and D. M. Hudgins. "From Interstellar Polycyclic Aromatic Hydrocarbons and Ice to Astrobiology." In *Solid State Astrochemistry*, edited by V. Pirronello, J. Krelowski, and Giulio Manicò, 251–316. NATO Science Series II: Mathematics, Physics and Chemistry. Dordrecht, Netherlands: Kluwer Academic, 2003.

Amato, P., M. Parazols, M. Sancelme, P. Laj, G. Mailhot, and A.-M. Delort. "Microorganisms Isolated from the Water Phase of Tropospheric Clouds at the Puy de Dôme: Major Groups and Growth Abilities at Low Temperatures." *FEMS Microbiology Ecology* 59 (2006): 242–54.

Angier, Natalie. *The Canon: A Whirligig Tour of the Beautiful Basics of Science*. New York: Houghton Mifflin, 2007.

Asaravala, Amit. "Are Nanobacteria Making Us Ill?" *Wired*, March 14, 2005, http://www.wired.com/science/discoveries.

Bains, William. "Many Chemistries Could Be Used to Build Living Systems." *Astrobiology* 4, no. 2 (2004): 137–67.

Bandea, Claudiu I. "The Origin and Evolution of Viruses as Molecular Organisms." *Nature Precedings*, October 23, 2009. http://precedings.nature.com/documents/3886/version/1.

Barlowe, Wayne Douglas, Ian Summers, and Beth Meacham. *Barlowe's Guide to Extraterrestrials: Great Aliens from Science Fiction Literature*. 2nd ed. New York: Workman, 1987.

Barrow, John D. *The Constants of Nature: From Alpha to Omega—The Numbers That Encode the Deepest Secrets of the Universe*. New York: Pantheon, 2002.

———. *Pi in the Sky: Counting, Thinking, and Being*. Oxford: Clarendon, 1992.

Barrow, John D., and Frank J. Tipler. *The Anthropic Cosmological Principle*. Oxford: Clarendon, 1986.

Belozerskaya, T., K. Aslanidi, A. Ivanova, N. Gessler, A. Egorova, Y. Karpenko, and S. Olishevskaya. "Characteristics of Extremophylic Fungi from Chernobyl Nuclear Power Plant." In *Current Research, Technology and Education Topics in Applied Microbiology and Microbial Biotechnology*, edited by A. Mendez-Vilaz, 88–94. Badajoz, Spain: Formatex Research Center, 2010.

Benner, S. A., A. Ricardo, and M. A. Carrigan. "Is There a Common Chemical Model for Life in the Universe?" *Current Opinion in Chemical Biology* 8 (2004): 672–89.

Borges, Jorge Luis. *Ficciones*. Edited by Anthony Kerrigan. New York: Grove, 1962.

Bradbury, Ray, Arthur C. Clarke, Bruce Murray, Carl Sagan, and Walter Sullivan. *Mars and the Mind of Man*. New York: Harper & Row, 1973.

Brock, Thomas Dale. "The Road to Yellowstone—and Beyond." *Annual Review of Microbiology* 49 (1995): 1–28.

Bruno, Giordano. *De l'infinito universo e mondi*. Translated by Dorothea Singer in *Giordano Bruno: His Life and Thought, with an Annotated Translation of His Work on the Infinite Universe and Worlds*. New York: Henry Schuman, 1950. Originally published in 1584.

Bryson, Bill. *A Short History of Nearly Everything*. New York: Broadway Books, 2003.

Bucher, M. A., and D. N. Spergel. "Inflation in a Low-Density Universe." *Scientific American*, January 1999.

Cano, R. J., and M. Borucki. "Revival and Identification of Spores in 25 to 40 Million Year Old Amber." *Science* 268 (1995): 1060–64.

Carter, Brandon. "The Anthropic Principle and Its Implications for Biological Evolution." *Philosophical Transactions of the Royal Society of London. A: Mathematical, Physical & Engineering Sciences* 310 (1983): 347.

Chapelle, F. H., K. O'Neill, P. M. Bradley, B. A. Methé, S. A. Ciufo, L. L. Knobel, and D. R. Lovley. "A Hydrogen-Based Subsurface Community Dominated by Methanogens." *Nature* 415 (2002): 312–16.

Chung, Daphne. "How Bug Extends Temperature Limit for Life." *New Scientist*, August 14, 2003.

Clark, Stephen. "ESA Chooses Jupiter as Destination for Science Probe." *Spaceflight Now*, May 2, 2012. http://www.spaceflightnow.com/news/n1205/02juice.

Clarke, A. C. *2001: A Space Odyssey*. New York: New American Library, 1968.

Cleland, Carol E. "Life without Definitions." *Synthese* 185 (2012): 125–44.

Cocconi, G., and P. Morrison. "Searching for Interstellar Communications." *Nature* 184 (1959): 844.

Cody, G. "Transition Metal Sulfides and the Origin of Metabolism." *Annual Review of Earth and Planetary Sciences* 32 (2004): 569–99.

Cooper, Henry S. F. *The Search for Life on Mars*. New York: Holt, Rinehart and Winston, 1980.

Coustenis, A., J. Lunine, D. Matson, C. Hansen, K. Reh, P. Beauchamp, J.-P. Lebreton, and C. Erd. "The Joint NASA-ESA Titan Saturn System Mission (TSSM) Study." Paper presented at the 40th Lunar and Planetary Science Conference, The Woodlands, TX, March 23–27, 2009. http://www.lpi.usra.edu/meetings/lpsc2009/pdf/1060.pdf.

Darwin, Charles. *The Life and Letters of Charles Darwin*, Vol. 2. London: Echo Library, 2007.

———. *The Origin of Species*. Introduction by George Levine. New York: Barnes & Noble, 2003.

———. *The Voyage of the Beagle: Journal of Researches into the Natural History and Geology of the Countries Visited during the Voyage of H.M.S. Beagle Round the World*. Introduction by Steve Jones. New York: Modern Library, 2001.

Davies, Paul. *The Eerie Silence: Renewing Our Search for Alien Intelligence*. New York: Houghton Mifflin Harcourt, 2010.

———. *The Goldilocks Enigma: Why Is the Universe Just Right for Life?* New York: Houghton Mifflin, 2006.

Davies, Paul C. W., Steven A. Benner, Carol E. Cleland, Charles H. Lineweaver, Christopher P. McKay, and Felisa Wolfe-Simon. "Signatures of a Shadow Biosphere." *Astrobiology* 9, no. 2 (2009): 241–49.

Dawkins, Richard. *The Selfish Gene*. New York: Oxford University Press, 1976.

de Duve, Christian. *A Guided Tour of the Living Cell*. 2 vols. New York: Scientific American/Rockefeller University Press, 1992.

Dick, Steven J. *The Biological Universe: The Twentieth-Century Extraterrestrial Life Debate and the Limits of Science*. New York: Cambridge University Press, 1996.

———. *Life on Other Worlds: The 20th-Century Extraterrestrial Life Debate*. Cambridge: Cambridge University Press, 2001.

Dickens, Charles. *Bleak House*. London: Bradbury and Evans, 1853.

Dobson, C. M., G. B. Ellison, A. F. Tuck, and V. Vaida. "Atmospheric Aerosols Are Prebiotic Chemical Reactors." *Proceedings of the National Academy of Sciences* 97 (2000): 11864–68.

Drake, Frank, and Dava Sobel. *Is Anyone Out There? The Scientific Search for Extraterrestrial Intelligence*. New York: Delacorte, 1992.

Dudzinski, Kathleen, and Toni Frohoff. *Dolphin Mysteries: Unlocking the Secrets of Communication*. New Haven, CT: Yale University Press, 2008.

Dyson, Freeman J. *Origins of Life*. Cambridge: Cambridge University Press, 1985.

——. "Time without End: Physics and Biology in an Open Universe." *Reviews of Modern Physics* 51, no. 3 (1979): 447–60.

Edmond, John M., and Karen Von Damm. "Hot Springs on the Ocean Floor." *Scientific American* (April 1983). doi:10.1038/scientific american0483-78.

Erwin, Terry L. "Tropical Forests: Their Richness in Coleoptera and Other Arthropod Species." *Coleopterists Bulletin* 36, no. 1 (1982): 74–75. doi:10.2307/4007977.

Everett, Hugh. "'Relative State' Formulation of Quantum Mechanics." *Reviews of Modern Physics* 29 (1957): 454–62.

Feinburg, G., and R. Shapiro. *Life beyond Earth: The Intelligent Earthling's Guide to Life in the Universe*. New York: William Morrow, 1980.

Folger, Tim. "Physics' Best Kept Secret." *Discover*, September 2001. http://discovermagazine.com/2001/sep/cover.

Folk, R. L. "SEM Imaging of Bacteria and Nanobacteria in Carbonate Sediments and Rocks." *Journal of Sedimentary Petrology* 63 (1993): 990.

Forward, Robert. "When You Live upon a Star." *New Scientist* 36 (December 24/31, 1987): 36–38.

Fountain, Henry. "Date Palm Seed from Masada Sprouts." *New York Times*, June 17, 2008.

Fox, Douglas. "Life—but Not as We Know It." *New Scientist* 194 (June 9, 2007): 34–39.

Freudenthal, Hans. *Lincos: Design of a Language for Cosmic Intercourse*. Amsterdam: North-Holland, 1960.

Funch, Peter, and Reinhardt Møbjerg Kristensen. "Cycliophora Is a New Phylum with Affinities to Entoprocta and Ectoprocta." *Nature* 378 (1995): 711–14. doi:10.1038/378711a0.

Gardner, Martin. "The Fantastic Combinations of John Conway's New Solitaire Game 'Life.'" *Scientific American*, October 1970.

Gaudiosi, John. "The 'Sims' Designer Creating New Game for Real

Life." *Reuters Canada*, January 2, 2012. http://ca.reuters.com/article/entertainmentNews/idCATRE8010L020120102?sp=true.

Gopnik, Adam. "Plant TV." *New Yorker*, March 15, 2010.

Greene, Brian. *The Hidden Reality: Parallel Universes and the Deep Laws of the Cosmos*. New York: Knopf, 2011.

Gribbin, John, and Mary Gribbin. *James Lovelock: In Search of Gaia*. Princeton, NJ: Princeton University Press, 2009.

Gross, Michael. *Life on the Edge: Amazing Creatures Thriving in Extreme Environments*. New York: Basic Books, 2001.

Gugliotta, Guy. "Fountains of Optimism for Life Way Out There." *New York Times*, May 9, 2011.

Harnik, Roni, Graham D. Kribs, and Gilad Perez. "A Universe without Weak Interactions." *Physical Review D* 74, no. 3 (2006): 035006-1–035006-15.

Harrison, Edward. *Cosmology: The Science of the Universe*. 2nd ed. Cambridge: Cambridge University Press, 2000.

———. "The Natural Selection of Universes Containing Intelligent Life." *Quarterly Journal of the Royal Astronomical Society* 36, no. 193 (1995): 193–203.

Hart, Michael H. "An Explanation for the Absence of Extraterrestrials on Earth." *Quarterly Journal of the Royal Astronomical Society* 16 (1975): 128–35.

———. "Habitable Zones about Main Sequence Stars." *Icarus* 37 (1979): 351–57.

Hart, Michael H., and Ben Zimmerman, eds. *Extraterrestrials— Where Are They?* 2nd ed. Cambridge: Cambridge University Press, 1995.

Hart, Stephen. "Hydrothermal Vents—Life's First Home?" SpaceRef, Ames Research Center, November 8, 2001. http://www.spaceref.com/news/viewpr.html?pid=6530.

Hempel, Amy, and Jim Shepard. *Unleashed: Poems by Writers' Dogs*. New York: Three Rivers Press, 1999.

Hooke, Robert. *Micrographia: or, Some Physiological Descriptions of Minute Bodies Made by Magnifying Glasses with Observations and Inquiries Thereupon.* London: Printed for James Allestry, 1667.

Horowitz, N. H. "The Biological Significance of the Search for Extraterrestrial Life." In *Advances in the Astronautical Sciences*, Vol. 22: *The Search for Extraterrestrial Life*, ed. J. S. Hanrahan, 3–13. Sun Valley, CA: Scholarly Publ., 1967.

Hoyle, Fred. *The Black Cloud.* New York: New American Library, 1959.

———. "The Universe: Past and Present Reflections." *Engineering and Science*, November 1981, 8–12.

Hussmann, Hauke, Frank Sohl, and Tilman Spohn. "Subsurface Oceans and Deep Interiors of Medium-Sized Outer Planet Satellites and Large Trans-Neptunian Objects." *Icarus* 185 (2006): 258–73.

Huxley, T. H. "On the Physical Basis of Life." In *Collected Essays*, Vol. 1: *Method and Results*, 130–65. New York: Greenwood Press, 1968.

Huygens, Christiaan. *The Celestial Worlds Discovered, or Conjectures concerning the Inhabitants, Plants and Productions of the World in the Planets.* 2nd ed. London: Printed for James Knapton, 1722.

Impey, Chris. *The Living Cosmos: Our Search for Life in the Universe.* New York: Random House, 2007.

———. "The New Habitable Zones." *Sky & Telescope*, October 2009.

Imshenetsky, A. A., S. V. Lysenko, and G. A. Kazakov. "Upper Boundary of the Biosphere." *Applied and Environmental Microbiology* 35 (1978): 1–5.

Islam, Jamal N. "Possible Ultimate Fate of the Universe." *Quarterly Journal of the Royal Astronomical Society* 18 (1977): 3–8.

Jaffe, Robert L., Alejandro Jenkins, and Itamar Kimchi. "Quark Masses: An Environmental Impact Statement." *Physical Review D* 79, no. 6 (2009): 065014-1–065014-33.

Jenkins, Alejandro, and Gilad Perez. "Looking for Life in the Multiverse." *Scientific American*, January 2010.

Kajander, E. O., and N. Ciftcioglu. "Nanobacteria: An Alternative Mechanism for Pathogenic Intra- and Extracellular Calcification and Stone Formation." *Proceedings of the National Academy of Sciences* 95 (1998): 8274.

Kanellos, Michael. "Moore's Law to Roll On for Another Decade." *CNET News*, February 10, 2003. http://news.cnet.com/2100-1001 -984051.html.

Kaufman, Marc. "Second Genesis on Earth." *Washington Post*, December 2, 2010.

Kelvin, William Thomson. "On the Origin of Life" (excerpt from the Presidential Address to the British Association for the Advancement of Science; held at Edinburgh in August, 1871). Reprinted in Kelvin's *Popular Lectures and Addresses*, Vol. 2: *Geology and General Physics*. London: Macmillan, 1894.

Keosian, John. *The Origin of Life*. 2nd ed. New York: Reinhold, 1968.

Kidd, John. *On the Adaptation of External Nature to the Physical Condition of Man: Principally with Reference to the Supply of His Wants and the Exercise of His Intellectual Faculties*. London: William Pickering, 1834.

Kuhn, Thomas. *The Structure of Scientific Revolutions*. 3rd ed. Chicago: University of Chicago Press, 1996.

Lee, Henry. *The Vegetable Lamb of Tartary*. London: Sampson Low, Marston, Searle, & Rivington, 1887.

Lewis, David. *On the Plurality of Worlds*. Oxford: Blackwell, 1986.

"Life in the Universe." In *Project Cyclops: A Design Study of a System for Detecting Extraterrestrial Intelligent Life*, 3–28. NASA-CR-114445. Prepared under Stamford/NASA/Ames Research Center 1971 Summer Faculty Fellowship Program in Engineering Systems Design. [1971]. http://seti.berkeley.edu/sites/default/files/19730010095_1973010095.pdf.

Lin, L.-H., J. Hall, T. Onstott, T. Gihring, B. Lollar, E. Boice, L. Pratt, J. Lippmann-Pipke, and R. Bellamy. "Planktonic Microbial Communities Associated with Fracture-Derived Groundwater in a Deep Gold Mine of South Africa." *Geomicrobiology Journal* 23 (2006): 475–97.

Lunine, Jonathan I. "Saturn's Titan: A Strict Test for Life's Cosmic

Ubiquity." *Proceedings of the American Philosophical Society* 153 (2009): 404–19.

———. "Titan as an Analog of Earth's Past and Future." *European Physical Journal. Conferences* 1 (2009): 267–74.

MacElroy, R. D. "Some Comments on the Evolution of Extremophiles." *Biosystems* 6 (1974): 74–75.

Maher, K. A., and D. J. Stevenson. "Impact Frustration of the Origin of Life." *Nature* 331 (1988): 612–14.

Margulis, Lynn, and Dorion Sagan. *What Is Life?* Foreword by Niles Eldredge. Berkeley: University of California Press, 1995.

Marine Biological Laboratory. "Life at the Extremes" (press release), September 24, 2001. http://www.mbl.edu/news/press_releases/2001/2001_pr_9_24_01.html.

Maude, A. D. "Life in the Sun." In *The Scientist Speculates*, edited by I. J. Good, 240–46. New York: Basic Books, 1963.

McConnell, Brian S. *Beyond Contact: A Guide to SETI and Communicating with Alien Civilizations*. Sebastopol, CA: O'Reilly Media, 2001.

McKay, C. P., and H. D. Smith. "Possibilities for Methanogenic Life in Liquid Methane on the Surface of Titan." *Icarus* 178 (2005): 274–76.

Minsky, Marvin. "Will Robots Inherit the Earth?" *Scientific American*, October 1994.

Mora, Camilo, Derek P. Tittensor, Sina Adl, Alastair G. B. Simpson, and Boris Worm. "How Many Species Are There on Earth and in the Ocean?" *PLoS Biology* 9, no. 8 (2011): e1001127. doi:10.1371/journal.pbio.1001127.

Moreira, D., and P. López-García. "Ten Reasons to Exclude Viruses from the Tree of Life." *Nature Reviews Microbiology* 7 (2009): 306–11.

Morowitz, Harold J. *Beginnings of Cellular Life: Metabolism Recapitulates Biogenesis*. New Haven, CT: Yale University Press, 1992.

Morowitz, Harold, and Carl Sagan. "Life in the Clouds of Venus?" *Nature* 215 (1967): 1259–60.

Morrison, Philip. "Twenty-Five Years of the Search for Extraterrestrial Communications." In *The Search for Extraterrestrial Life—Recent Developments: Proceedings of the 112th Symposium of the International Astronomical Union Held at Boston University, Boston, Mass., USA, June 18–21, 1984*, edited by Michael D. Papagiannis, 13–19. Dordrecht, Netherlands: Kluwer Academic, 1985.

Mullen, Leslie. "Swimming in a Salty Sea." *Astrobiology Magazine*, November 19, 2007. http://www.astrobio.net/exclusive/2528/swimming-a-salty-sea.

Muller, T., W. Zilche, and N. Auner. "Recent Advances in the Chemistry of Si-Heteroatom Multiple Bonds." In *The Chemistry of Organic Silicon Compounds*, Vol. 2, part 1, edited by Z. Rappoport and Y. Apeloig, 857–1062. Chichester, UK: Wiley, 1998.

Murphy, M. T., J. K. Webb, V. V. Flambaum, V. A. Dzuba, C. W. Churchill, J. X. Prochaska, J. D. Barrow, and A. M. Wolfe. "Possible Evidence for a Variable Fine-Structure Constant from QSO Absorption Lines: Motivations, Analysis and Results." *Monthly Notices of the Royal Astronomical Society* 327 (2001): 1208.

Nagel, Thomas. "What Is It Like to Be a Bat?" *Philosophical Review* 83 (1974): 435–50.

National Research Council. Committee on the Limits of Organic Life in Planetary Systems. *The Limits of Organic Life in Planetary Systems*. Washington, DC: National Academies Press, 2007.

Navarro-González, R., F. A. Rainey, P. Molina, D. R. Bagaley, B. J. Hollen, J. de la Rosa, A. M. Small, et al. "Mars-Like Soils in the Atacama Desert, Chile, and the Dry Limit of Microbial Life." *Science* 302 (2003): 1018–21.

Nevalla, Amy E. "On the Seafloor, a Parade of Roses." *Oceanus Magazine*, June 25, 2005.

Nozick, Robert. *Philosophical Explanations*. Cambridge, MA: Harvard University Press, 1981.

Nuland, Sherwin. *How We Live: The Wisdom of the Body*. London: Vintage, 1998.

Pace, Norman R. "A Molecular View of Microbial Diversity and the Biosphere." *Science* 276 (1997): 734–40.

——. "The Universal Nature of Biochemistry." *Proceedings of the National Academy of Sciences* 98 (2001): 805–8.

Pedersen, K. "Exploration of Deep Intraterrestrial Microbial Life: Current Perspectives." *FEMS Microbiology Letters* 185 (2000): 9–16.

Piccard, Jacques, and Robert Sinclair Dietz. *Seven Miles Down: The Story of the Bathyscaph Trieste.* New York: Putnam, 1961.

Pikuta, E. V., R. B. Hoover, B. Klyce, P. C. W. Davies, and P. Davies. "Bacterial Utilization of *L*-Sugars and *D*-Amino Acids," *Proceedings of SPIE* 6309 (2006). http://dx.doi.org/10.1117/12.690434.

Pirie, N. W. "The Meaninglessness of the Terms 'Life' and 'Living,'" in *Perspectives in Biochemistry*, edited by J. Needham and D. R. Green, 11–22. Cambridge: [Cambridge] University Press, 1937.

Pittendrigh, C. S., Wolf Vishniac, and J. P. T. Pearman. *Biology and the Exploration of Mars: Report of a Study.* Washington, DC: National Academy of Sciences–National Research Council, 1966.

Reaves, M. L., S. Sinha, J. D. Rabinowitz, L. Kruglyak, and R. J. Redfield. "Absence of Arsenate in DNA from Arsenate-Grown GFAJ-1 Cells." Submitted to *Science* on January 31, 2012. arXiv:1201.6643v1.

Reyes-Ruiz, M., C. E. Chavez, M. S. Hernandez, R. Vazquez, H. Aceves, and P. G. Nuñez. "Dynamics of Escaping Earth Ejecta and Their Collision Probability with Different Solar System Bodies." Submitted to *Icarus* on August 17, 2011. arXiv:1108.3375v1.

Robinson, Margaret W. *Fictitious Beasts: A Bibliography.* London: Library Association, 1961.

Rothman, Tony. "A 'What You See Is What You Beget' Theory." *Discover*, May 1987.

Rothschild, Lynn J., and Rocco L. Mancinelli. "Life in Extreme Environments." *Nature* 409 (2001), 1097.

Roussel, Erwan G., Marie-Anne Cambon Bonavita, Joël Querellou, Barry A. Cragg, Gordon Webster, Daniel Prieur, and R. John

Parkes. "Extending the Sub-Sea-Floor Biosphere." *Science* 320 (2008): 1046.

Sagan, Carl. *Cosmos.* New York: Random House, 1980.

Sagan, Carl, and E. E. Saltpeter. "Particles, Environments, and Possible Ecologies in the Jovian Atmosphere." *Astrophysical Journal Supplement Series* 32 (1976): 737–55.

Sattler, B., H. Puxbaum, and R. Psenner. "Bacterial Growth in Supercooled Cloud Droplets." *Geophysical Research Letters* 28 (2001): 239–42.

Schenk, Paul M., Clark R. Chapman, Kevin Zahnle, and Jeffrey M. Moore. "Ages and Interiors: The Cratering Record of the Galilean Satellite." In *Jupiter: The Planet, Satellites and Magnetosphere*, edited by Fran Bagenal, Timothy E. Dowling, and William B. McKinnon, 427–56. Cambridge: Cambridge University Press, 2004.

Scheraga, H. A., M. Khalili, and A. Liwo. "Protein-Folding Dynamics: Overview of Molecular Simulation Techniques." *Annual Review of Physical Chemistry* 58 (2007): 57–83.

Schieber, Jürgen, and Howard J. Arnott. "Nannobacteria as a By-Product of Enzyme-Driven Tissue Decay." *Geology* 31 (August 2003): 717–20.

Schrödinger, Erwin. *What Is Life? The Physical Aspect of the Living Cell.* Cambridge: [Cambridge] University Press, 1944.

Schulze-Makuch, D., and D. H. Grinspoon. "Biologically Enhanced Energy and Carbon Cycling on Titan?" *Astrobiology* 5 (2005): 560–64.

Schulze-Makuch, Dirk, David H. Grinspoon, Ousama Abbas, Louis N. Irwin, and Mark A. Bullock. "A Sulfur-Based Survival Strategy for Putative Phototrophic Life in the Venusian Atmosphere." *Astrobiology* 4 (2004): 11–18.

Schulze-Makuch, D., and L. N. Irwin. "Reassessing the Possibility of Life on Venus: Proposal for an Astrobiology Mission." *Astrobiology* 2 (2002): 197–202.

SETI League, "Declaration of Principles Concerning Activities Following the Detection of Extraterrestrial Intelligence," last updated January 4, 2003, http://www.setileague.org/general/protocol.htm.

Seuss, Dr. *If I Ran the Zoo*. New York: Random House, 1950.

Shiga, David. "Hints of Life Found on Saturn's Moon." *New Scientist*, June 4, 2010.

———. "NASA Floats Titan Boat Concept." *New Scientist*, May 9, 2011. http://www.newscientist.com/article/dn20459-nasa-floats-titan-boat-concept.html.

Shklovskii, I. S., and Carl Sagan. *Intelligent Life in the Universe*. Boca Raton, FL: Emerson-Adams, 1998.

Shostak, Seth. *Confessions of an Alien Hunter: A Scientist's Search for Extraterrestrial Intelligence*. Foreword by Frank Drake. Washington, DC: National Geographic Society, 2009.

"Size Limits of Very Small Microorganisms: Proceedings of a Workshop." Washington, DC: National Academy Press, 1999.

Slater, A. E. "Biological Problems of Space Fight: A Report of Professor Haldane's Lecture to the Society on April 7, 1951." *Journal of the British Interplanetary Society* 10 (1951): 154–58.

Smith, Maurice. "Nanobes." *Micscape Magazine*, March 1999.

Sogin, Mitch. "In Search of Diversity." *Astrobiology Magazine*, June 20, 2005. http://www.astrobio.net/index.php?option=com_retro spection&task=detail&id=1608&fid=28&pid=5.

Spohn, Tilman, and Gerald Schubert. "Oceans in the Icy Galilean Satellites of Jupiter?" *Icarus* 161 (2003): 456–67.

Stanier, R. Y., M. Dondoroff, and E. A. Adelberg. *The Microbial World*. Englewood Cliffs, NJ: Prentice-Hall, 1957.

Stetter, Karl O. "Hyperthermophilic Procaryotes." *FEMS Microbiology Reviews* 18 (1996): 149–58.

Stevens, T. O., and J. P. McKinley. "Lithoautotrophic Microbial Ecosystems in Deep Basalt Aquifers." *Science* 270 (1995): 450–54.

Strobel, D. F. "Molecular Hydrogen in Titan's Atmosphere: Implications of the Measured Tropospheric and Thermospheric Mole Fractions." *Icarus* 208 (2010): 878–86. doi:10.1016/j.icarus.2010.03 .003.

Tegmark, Max. "The Multiverse Hierarchy." Submitted on May 8, 2009. arXiv:0905.1283v1.

———. "Parallel Universes." *Scientific American*, May 2003.

Tenenbaum, David. "Making Sense of Mars Methane." *Astrobiology Magazine*, June 9, 2008. http://astrobiology.nasa.gov/articles/making-sense-of-mars-methane.

Tsytovich, V. N., G. E. Mook

rfill, V. E. Fortov, N. G. Gusein-Zade, B. A. Klumov, and S. V. Vladimirov. "From Plasma Crystals and Helical Structures towards Inorganic Living Matter." *New Journal of Physics* 9 (2007): 263.

Uwins, P. J. R., R. I. Webb, and A. P. Taylor. "Novel Nano-organisms from Australian Sandstones." *American Mineralogist* 83 (1998): 1541–50.

Ventosa, A., D. R. Arahal, and B. E. Volcani. "Studies on the Microbiota of the Dead Sea—50 Years Later." In *Microbiology and Biogeochemistry of Hypersaline Environments*, edited by A. Oren, 139–47. Boca Raton, FL: CRC Press, 1999.

Vernier, J. P. "The SF of J. H. Rosny the Elder." *Science Fiction Studies* 2, part 2 (no. 6), July 1975. http://www.depauw.edu/sfs/back issues/6/vernier6art.htm.

Viewing, David. "Directly Interacting Extra-terrestrial Technological Communities." *Journal of the British Interplanetary Society* 28 (1975): 735–44.

Vinge, Vernor. "The Coming Technological Singularity: How to Survive in the Post-Human Era." Reprint of address delivered at the VISION-21 Symposium sponsored by NASA Lewis Research Center and the Ohio Aerospace Institute, March 30–31, 1993. http://www-rohan.sdsu.edu/faculty/vinge/misc/singularity.html.

Vreeland, R. H., W. D. Rosenzweig, and D. W. Powers. "Isolation of a 250-Million Year Old Halotolerant Bacterium from a Primary Salt Crystal." *Nature* 407 (2000): 897–900.

Wallace, A. R. *Man's Place in the Universe: A Study of the Results of Scientific Research in Relation to the Unity or Plurality of Worlds.* 4th ed. London: George Bell, 1904.

Wells, H. G. *H. G. Wells: Early Writings in Science and Science Fiction*. Edited by Robert M. Philmus and David Y. Hughes. Berkeley: University of California Press, 1975.

West, Anthony. *H. G. Wells: Aspects of a Life*. New York: Random House, 1984.

Wharton, D. A., and D. J. Ferns. "Survival of Intracelluar Freezing by the Antarctic Nematode *Panagrolaimus davidi*." *Journal of Experimental Biology* 198 (1995): 1381–87.

Whitrow, Gerald. *The Structure and Evolution of the Universe: An Introduction to Cosmology*. New York: Harper, 1959.

———. "Why Space Has Three Dimensions." *British Journal for the Philosophy of Science* 6, no. 21 (1955): 23–24.

Wilson, Edward O. *Anthill: A Novel*. New York: W. W. Norton, 2010.

———. *The Diversity of Life*. New York: W. W. Norton, 1999.

———. *The Future of Life*. New York: Knopf, 2002.

Wolfe-Simon, Felisa, Jodi Switzer Blum, Thomas R. Kulp, Gwyneth W. Gordon, Shelley E. Hoeft, Jennifer Pett-Ridge, John F. Stolz, et al. "A Bacterium That Can Grow by Using Arsenic Instead of Phosphorus." *Science Express*, December 1, 2010.

Young, John D., and Jan Martel. "The Truth about Nanobacteria [Preview]." *Scientific American*, January 2010. doi:10.1038/scientificamerican0110-52.

Zimmer, Carl. "This Paper Should Not Have Been Published." *Slate*, December 7, 2010. http://www.slate.com/articles/health_and_science/science/2010/12/this_paper_should_not_have_been_published.html.

Index

INDEX

Barcode of Life project, 39
Barlowe's Guide to Extraterrestrials, 165
barophiles (pressure lovers), 15
Barrow, John D., 190, 206, 212
bathyscaphes, 8
bathyspheres, 8, 117
Baxter, Stephen, 168*n*
Benford, Gregory, 168*n*, 174*n*
Benner, Steven, 49–50, 57, 107
Bentley, Richard, 133
Bernal, John Desmond, 130
bestiaries, vii–xi
   dinosaur, x–xi
   fictitious, vii–ix
   microbial, ix–x
biochemistry, xii, xv, 64, 65*n*, 86–87, 88,
     89–93, 96, 98, 112, 144, 155, 202, 216,
     219–20
   "ammoniated," 101, 152
   in science fiction, 166, 167, 171, 173*n*, 177
   silicon-based, 89–91, 152, 167, 168–69,
     173*n*–74*n*
   of Triton, 116
biodiversity, viii, 39, 43, 233*n*
biosignatures, 44–45, 78, 83, 219
biosolvents, 107, 167, 168*n*
Black Cloud, The (Hoyle), 172–73, 175
black holes, 132, 134, 135–36, 182, 215
   event horizons of, 136, 193
   as habitats, 168, 174*n*, 216
black smokers, 10–11
Bleak House (Dickens), xi*n*
Borges, Jorge Luis, 206*n*
Bostrom, Nick, 212, 214*n*
Bradbury, Ray, 64*n*
Brin, David, 168*n*
brine inclusions, 25, 230*n*
brine shrimp (*Artemia salina*), 24–25
Brock, Thomas Dale, 11–14
brown dwarf stars, 134
Browning, Elizabeth Barrett and Robert, 160
Bruno, Giordano, 180
Burroughs, Edgar Rice, 64*n*

California redwood trees (*Sequoiadendron*
     giganteum and *Sequoia sempervirens*),
     18–19
carbon, 1, 21*n*, 35, 46, 51, 88, 90–93, 94,
     95–96, 106–7, 118, 125, 135, 199–202
Carter, Brandon, 113, 186, 187–91
Cassini spacecraft, 102–10

cell membranes, 19–20, 21, 22, 24, 30, 88,
     95, 128
cells, xiv, 18, 19–21, 28, 33, 46, 56, 86–88, 98,
     216, 230*n*
   anhydrobiosis of, 24–27
   cytoplasm of, 19–20, 22–23, 30, 87
   division of, 87
   first appearance of, 140
   freezing of, 21
   mitochondria of, 43–44
   nuclei of, 87, 113
   phosphorus in, 51
   proteins in, 86–87
   protoplasm of, 89*n*–90*n*
   shared features of, 30
   size of, 56–59, 67
   vesicle precursors of, 128–29
cell wall, 19–20
Celsius temperature scale, 19
Census of Marine Life, 39
chemosynthesis, 9–10, 43
Chernobyl nuclear reactor, 28
chirality, 54–56
   mirror, 55–56
chromosomes, xi, 28
Cisar, John, 58
Clarke, Arthur C., 154, 168*n*
Cleland, Carol, 34, 42, 45, 59, 60, 71
Clement, Hal, 173–75
closed-basin lakes, 4, 22–23
   Mono Lake, 52–54, 56
clouds, 32
   interstellar molecular, 125–26, 127, 129,
     132
   microbes in, xiv, 117–18
   sentient interstellar, 172–73, 175, 189, 221
   Venusian, 117–19
Cocconi, Giuseppe, 138–39, 147
cold lovers (cryophiles; psychrophiles), 15, 16
   habitats for, 93–94, 97–98
Cold War, 6
Columbia River, 42
comets, 98*n*, 126, 127–30
Committee on the Limits of Organic Life in
     Planetary Systems, xvii
Communications Research Institute, 161–62
continental drift, 3, 4–11
convergent evolution, 46–47
Conway, John, 69, 209
Coon, Gene L., 168
Copely, Shelley, 34, 55

INDEX

Le Guin, Ursula, 166
Lem, Stanislaw, 175–76
leptons, 187
Levin, Gilbert, 74, 75, 79, 80, 81, 82
Lewis, David, 206
life:
  ambisexual, 166
  chemical composition of, 50–51, 125
  complex, relatively late appearance of,
    140–41
  energy source of, 84, 86, 95, 96, 113
  extraterrestrial, *see* extraterrestrial life
  first appearance of, 113, 141, 150
  limits of, xii–xvii, 21–22, 31
  liquid medium needed by, 84–86, 96–98
  liquid water and, xiii, 16–19, 63–65, 85,
    109, 114
  multicellular, 113
  need for second example of, 71
  needs of, 83–88, 89, 95–98
  nonlife vs., 65
  probability of, 33, 34
  reproductive process of, 87
  second genesis of, 33, 35–38, 47, 52, 55
  semipermeable barrier needed by, 87–88,
    95
  theory of, 71–75
  trace elements in, $51n$
  at very cold temperatures, 93–94, 97–98
life, defining, xii–xiii, xvii, 47, 61–82, 83–84,
    209–10, 215
  by each biology specialty, 64–65
  evolution as, 67–71, 83, 209
  functions in, 65, 66, 67
  materialist view in, 66
  nonlife and, 65, 67–69
  by philosophers, 65, 70, 71
  theory of life needed for, 71–75; *see also*
    Viking spacecraft, life-detecting
    experiments of
  vital force in, 66, $233n$–$34n$
life, origin of, 26–31, 32–33, 35–37, 47,
    123–24, 154
  Darwin on, 28, 129, $231n$–$32n$, $234n$
  extraterrestrial, 26–27, 33, 36–37, 129
  "last common ancestor" and, xv, xviii, 28,
    30–31, 33, $165n$, $231n$–$32n$
  on Mars, 37
  nonlife-to-life transition in, 47–48, 66
  prebiotic molecules and, 28, 51–52, 55,
    123–30

life detection experiments, 136–37
  see also Viking spacecraft, life-detecting
    experiments of
life sciences departments, academic, 14
Lilly, John C., 161–62
Limits of Organic Life on Planetary Systems,
    The (National Research Council), xvii,
    20–21, 22, 41, 49, 53, 64, 70, 106–7,
    209, 219, $240n$
Lindsay, David, 175
lipids, 88, 128
Lovelock, James, 40, 72–74, 78, 108–9
Lovley, Derek, 20
Lowell, Percival, $64n$
LUCA (last universal common ancestor),
    30–31, $165n$
  see also life, origin of
Lucian of Samosata, 133
Lunine, Jonathan, 113–14, 145, 147

MacElroy, R. D., 15, 16
machine-based life, 152–55
macromolecules, 88, 90, 95–96
magnesium sulfate, 101
Maher, Kevin, 36
Man and Dolphin (Lilly), 161
Manual of Systematic Bacteriology (Bergey),
    40
Marcy, Geoff, 144
Margulis, Lynn, 40, 73–74, $234n$
Mariner spacecraft, 77
Mars, xii–xiii, xvi, 16, 26
  atmosphere of, 37, $45n$, 61–62, 73, 75,
    $234n$–$35n$
  Chryse plain of, 78–79, 80, 99
  Earth material traded with, xvii, 37
  as extremophile habitat, 63–64
  habitable zone position of, 62, 96, 97
  landscape of, 78–79
  life as originating on, 37
  life on, 45, 63, 71, 72–82, $231n$; *see also*
    Viking spacecraft, life-detecting
    experiments of
  in science fiction, $64n$, 89, 177
  terraforming of, 64
  water on, 61–62, 75, 81–82
Mars Science Laboratory mission, 45,
    218–19
Martian meteorite ALH84001, 57
"Martian Odyssey, A" (Weinbaum), $173n$–$74n$
Masada, excavation of, 25

INDEX

INDEX

Volcani, Benjamin, 22–23, 231*n*
volcanoes, xiii
  hot springs heated by, 5
  seafloor, 2–3
  on Titan, 106
von Däniken, Erich, 147
Voyager spacecraft, 72, 77, 80, 115–16, 121–22
Voyage to Arcturus, A (Lindsay), 175

Wallace, Alfred Russel, xi, 186*n*
Walsh, Don, 8
War of the Worlds (Wells), 89
water, 16–23, 35, 84–85, 111, 114, 120, 126, 128
  acquisition and retention of, 18–22
  anhydrobiosis and, 24–27
  in brine inclusions, 25, 230*n*
  definition of, 71, 85
  as diffusion medium, 17
  on Europa, 100–101
  harmful effects of, 85*n*
  ice, xiii, 18, 21, 32, 42, 105, 230*n*
  life linked to, xiii, 16–19, 63–65, 85, 109, 114
  on Mars, 61–62, 75, 81–82
  molecular weight of, 88
  organic solvents in, 21–22
  phase transition of, 47
  polarized molecule of, 17–18
  salt, 22–23, 24–25, 101
  SETI searches and, 143, 145
  in Solar System, 61–65
  solutes of, 21–23
  as solvent, xv, 17, 21, 46, 88, 90–91, 118
  sulfuric acid in, 118
  surface tension of, 17, 85

  temperature levels of, xiii, xiv, 7, 10–11, 12, 17–18, 19–22, 42, 85, 94, 230*n*
  on Titan, 105, 106, 107
  on Venus, 119
Watson, James, xii
Wegner, Alfred, 4
Weinbaum, Stanley G., 173*n*–74*n*
Weinberg, Steven, 186, 197–98, 207
weird life:
  definition of, 165*n*
  early publications on, xvi
  extraterrestrial, xvi–xvii
  life's limits and, xiv–xvii
  as term, xvii–xviii
Wells, H. G., 64*n*, 89–90, 94, 154, 176, 177
West Mata volcano, 2–3
"What Is It Like to Be a Bat?" (Nagel), 160
What Is Life? (Schrödinger), 66, 70, 73, 234*n*
White, James, 169–70
white dwarf stars, 134, 135, 170, 215
Whitrow, Gerald, 207–8
"Will Robots Inherit the Earth?" (Minsky), 152
Wilson, Edward O., 39, 41, 161
Woese, Carl, 15–16, 53
Wolfe-Simon, Felisa, 52–54
Woods Hole Oceanographic Institution (WHOI), 1–3, 5–11
Woolf, Virginia, 160
Wright, Will, 214

Yellowstone National Park, 12–14

Zeeman effect, 139
Zettler, Linda Amaral, 23–24
zooplankton, 8